Sampling Inspection and Quality Control

Sampling Inspection and Quality Control

G. Barrie Wetherill

Professor of Statistics
University of Newcastle upon Tyne

SECOND EDITION

LONDON NEW YORK
CHAPMAN AND HALL

First published 1969
by Methuen & Co. Ltd
Second Edition 1977
published as a Science Paperback
by Chapman and Hall Ltd,
11 *New Fetter Lane, London EC4P 4EE*
Published in the USA by Chapman and Hall
29 *West* 35*th Street, New York NY* 10001
Reprinted 1982, 1986

Printed in Great Britain at the
University Press, Cambridge

ISBN 0 412 14960 5

Preface

Since the pioneering work of Dodge and Romig in the 1920's there has grown up a vast literature on sampling inspection and quality control. However, most of the available texts are written for personnel of inspection departments, giving practical details of exactly what should be done to operate various plans. Many of these are excellent books for their purpose and it is not my intention to attempt to replace them, and indeed I would not be qualified to do this.

My intention in this book has rather been to give a broad coverage of the field, with some emphasis on the principles upon which various plans are constructed. I have also given a simple treatment of important background theory. I hope that the book will be suitable for courses in Universities and Technical Colleges.

The lack of a book of this kind is partially responsible for many statisticians and operational research workers finishing their training with only a smattering of knowledge of this important practical field.

Those interested in pursuing the theoretical aspects will find adequate references throughout, and at the end of the book a list of papers for further study.

Exercises are provided at the end of most sections, and some of these which may give difficulty are marked with an asterisk.

I am grateful to a number of colleagues for detailed comments on an earlier draft of this book, and I mention particularly Mr A. F. Bissell, Dr G. E. G. Campling, Professor D. R. Cox, Mr W. D. Ewan, Mr W. A. Hay and Dr D. V. Hinkley.

G. B. W.

Preface to the second edition

The principal changes in this edition are that tables, nomograms, and explanation have been added throughout so that numerical exercises can be set, and the sections on acceptance sampling have been rewritten. In chapter 3 I have included both ($1{\cdot}96\sigma$, $3{\cdot}09\sigma$) and (2σ, 3σ) limits for control charts, and in chapter 4 I have included an explanation of the use of the nomogram for designing CUSUM schemes. In numerous places the text has been brought up to date with current work. I am indebted to Professor K. W. Kemp and the editors of *Applied Statistics* for permission to reproduce the CUSUM nomogram given in Appendix II.

G. B. W.

Contents

1. Introduction

1.1 Examples and definitions

The importance of sampling inspection and quality control procedures is very widely accepted, and there is a long history of applications to various branches of industry. The purpose of this book is to give a brief account of the procedures available, and to outline the principles upon which they are based. Some typical situations are illustrated in the following examples.

Example 1.1 (Griffiths and Rao, 1964). Large batches of electrical components have been purchased for manufacture into parts of a computer. Each batch contains an unknown proportion of defective components which will cause faults at a later stage if passed on to the manufacturing process. It was decided that any batch containing more than a critical proportion p_0 of defectives should be rejected, and that a single sampling plan was to be operated. This plan was to select n items from the batch at random, and reject the batch if the number of defective items found in the sample were greater than some quantity c. □□□

Example 1.2. Morgan *et al.* (1951) have described a sampling procedure used in the grading of milk. Films of milk were prepared on slides and viewed under a microscope. Several microscopic fields were observed on each film, and the number of bacterial clumps counted. The observations were used to estimate the density of bacterial clumps in the milk, and it was this latter quantity which determined the grade of the milk. □□□

Example 1.3. Grant and Leavenworth (1972, pp. 16–27) described in detail two situations of the following type. The output of a production process is a continuous series of items and the most important characteristic of each item can be described by a single measurement, such as length, strength, etc. If the production process is operating correctly the measurements on the items are approximately normally distributed with a certain mean and variance. A sample of five items is drawn from the process every hour and measurements made on each item. From the results it is required to decide whether the process is operating correctly (the term 'in control' is used), or whether some kind of corrective action needs to be taken. Sometimes in such applications it is also required to decide whether the current output should be passed, or whether it needs to be sorted and reprocessed, etc. □ □ □

Examples 1.1 and 1.2 illustrate what we call *sampling inspection*; in these examples it is necessary to decide what to do with a given quantity of material, e.g. to decide whether to accept or reject a batch of goods. We say that we wish to *sentence* batches of goods.

Example 1.3 describes the *quality control* situation, where the interest is more in controlling the production process than sentencing goods. Inevitably there are many situations where the aim is both to control a process and sentence goods, so that it is impossible to draw a clear boundary between sampling inspection and quality control.

Example 1.2 illustrates a case where, in comparison with Example 1.1 or perhaps Example 1.3 the observations are relatively expensive. We shall see that this leads to a rather different sampling plan being appropriate for Example 1.2.

There are certain common features to the three examples. In each case procedures are required by which we decide among a small number of possible courses of action, and in each case the procedures are to use a small sample of observations, and not, for example, inspection of every item. Now in industry it is sometimes necessary to defend inspection by samples against 100% inspection, and to explain why sample procedures are reliable. Clearly there are some situations in which 100% inspection is desired rather than sampling inspection, but such situations are infrequent. The reasons why sample methods are preferred are as follows:

(*i*) We never require absolutely accurate information about a batch or quantity of goods to be sentenced. Thus in Example 1.1 it would be sufficient to estimate the percentage of defective items in the batch to within $\frac{1}{2}$% or so. Complete inspection in Example 1.1 would be an unnecessary waste of time and labour, unless the aim is to *sort* all the items into good and bad. For the purpose of sentencing the batch, an estimate of the percentage defective is quite sufficient.

(*ii*) A point allied to (*i*) is that under the usual assumptions, the standard error of an estimate reduces as the number of observations increases, approximately as the reciprocal of the square root of the number of observations. Therefore in order to halve the standard error we must take four times as many observations. Beyond a certain point it is either impractical or not worth while achieving greater accuracy.

(*iii*) Even if the entire batch is inspected in Example 1.1 say, we still do not have an absolutely accurate estimate of the percentage defective *unless inspection is perfect.* In industrial situations inspection is very rarely perfect and Hill (1962) quotes a probability of 0·9 as being 'not unreasonable' for the probability of recognizing defects by visual inspection. Some experiments have indicated that if inspectors are faced with batches for 100% inspection, then the inspection tends to be less accurate than if sample methods are used.

(*iv*) In some cases, such as in Example 1.2, inspection is very costly and 100% inspection is obviously ruled out. One case of this is *destructive testing,* as in testing of artillery shells. Another case of costly inspection is when complicated laboratory analyses are involved.

One situation where 100% inspection is appropriate is when it can be arranged cheaply by some automatic device. More usually sample methods will be appropriate.

When sample methods are employed we shall usually make the assumption that sampling is *random*. Thus in Example 1.1 a sample should be taken in such a way that every item in the batch is equally likely to be taken. In practice this assumption is rarely satisfied and this has to be taken into account when drawing up a plan.

Sometimes it is possible to stratify the items to be sentenced, and use this to draw up a more efficient sample procedure. For example, in the transport of bottled goods in cartons, the bottles next to the face of the carton are more likely to be damaged than those in the interior. In this case it would be better to define two strata, one being those bottles next to a face of the carton, and the other stratum being the remainder. A procedure which sampled these strata separately would be more efficient than a straight random sample. To the author's knowledge, very little use has been made of this kind of device.

1.2 Where inspection?

In any industrial process there are a number of places where inspection can and should be carried out. Consider an industrial process which produces nominally constant output over long periods of time, pictured diagrammatically in Figure 1.1. The process may be producing continuously, as in nylon spinning, or small components may be produced at a high rate, as in light engineering, or production may be non-continuous, as, say, of petrol engines, pottery, etc. Any such process can be thought of in three parts; the input stage, where the raw materials are accepted for the process, the process itself, and the output stage, where the product is passed on for sale, or for use in the next stage. Sometimes a process can be thought of as

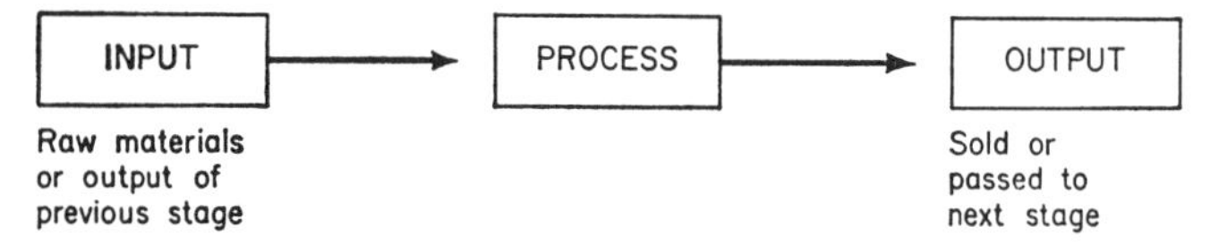

Figure 1.1. *A typical process.*

being composed of several stages, each as described in Figure 1.1, and the output of one stage is the input of the next, and often there are several inputs to a process. For example, car bodies are pressed and made in one factory, engines manufactured in another, tyres in a third, etc., and these are all inputs to the final stage of assembling finished cars.

We can now consider the inspection suitable for each of the three parts of the process featured in Figure 1.1.

INPUT. We may inspect the input to ensure that it is of sufficiently high quality. For example, in weaving cotton garments, yarn of low tensile strength leads to frequent breaks and loom stoppages. Bad material may be returned to the vendor, returned for reprocessing, scrapped, or set aside for a different use. If the quality offered at input is variable, sampling inspection here can save a good deal of trouble and money.

OUTPUT. We may inspect the output to reduce the risk of bad quality being passed on and causing loss of prestige and loss of money if bad quality items must be replaced. If there is a guarantee, the manufacturer will wish to guard against too many claims against this. With items such as packets of detergent, it may be necessary to reduce the risk of prosecution for selling underweight. Sometimes the aim of inspection of the output is to earn some quality seal, such as a British Standards Institution mark. Any output rejected may be scrapped, sold as inferior products, or completely sorted and the defective items rectified or replaced.

PROCESS. It is usually possible to inspect the process itself, sometimes at several points, and check up how it is working. Two different aims might be involved with such inspection. Firstly it may be possible to use the information to adjust the process and so reduce the amount of bad production. Secondly, it may be desirable to sort out the bad production and sort or return articles for reprocessing before further processing costs are incurred. When the main aim is to control the process we have the quality control situation. Frequently inspection of the process has both process control and product sorting in view. For example, in production of chocolate bars, inspection of the process, before wrapping, may lead both to adjustments to the process and also to sorting out underweight production for melting down and reprocessing.

When planning any particular inspection plan, it is important to bear in mind the various possibilities for inspection. Sometimes inspection effort is more worth while at one place than another. The type of inspection plan which is appropriate depends on the particular situation, and the aims in view.

Exercises 1.2

1. Decribe a production process with which you are familiar. Detail the places in which inspection plans could be operated, and describe the action taken on inspection results.

1.3 Classification of inspection plans

Any system of classifying inspection plans is unsatisfactory in that borderline categories exist. Nevertheless it will be found useful to have some classification system. We shall first list different inspection situations and then give alternative sampling plans.

(*a*) *Inspection situations*

(*i*) *Batch inspection* or *continuous production inspection*. Batch inspection occurs when we have items presented in, say, boxes, and it is desired to pass sentence on each box of items together, and not on each individual item. If on the other hand we have continuous nylon thread, or a production line of continuously produced small items such as chocolate bars and items are *not* treated in batches for sentencing, then we have continuous production inspection. The essential distinction is whether items are batched for inspection purposes or not; often with a continuous production process, items are batched for inspection purposes. With batch inspection there is no need for any order in the batches presented, although sometimes there is an order, and this information can be used, see below. Example 1.4 illustrates one of the earliest types of continuous sampling plans (CSP); batch inspection plans are illustrated later in this section.

Example 1.4. *Dodge plan.* At the outset inspect every item until i successive items are found free of defects. Then inspect every nth item until a defect is found when 100% inspection is restored. □ □ □

(*ii*) *Rectifying inspection* or *acceptance inspection.* If, say, batches of items are presented for sentencing, and the possible decisions are, say, accept or reject, or accept or sell at a reduced price, etc., we have acceptance inspection. Rectifying inspection occurs when one of the possible decisions is to sort out the bad items from a batch and adjust or rectify them, or else replace them. That is, with rectifying inspection, the proportion of defective items may be changed.

(*iii*) *Inspection by attributes* or *inspection by variables.* Inspection by attributes occurs when items are classified simply as effective or defective, or when mechanical parts are checked by go–not-go gauges. The opposite of this is inspection by variables when the result of inspection is a measurement of length, the voltage at which a voltage regulator works, etc. An intermediate classification between these is when items are graded. There is frequently a choice between inspection by attributes or by variables, and also a choice of the number of such variables inspected. The choice between these depends on the costs of inspection, the type of labour employed, and also on the assumptions which can be made about the probability distribution of the measured quantities.

(*b*) *Alternative sampling plans*

We shall be mainly concerned here with batch inspection plans. Example 1.4 illustrates a continuous production inspection plan, and other such plans will be described later. An intermediate situation occurs when items are batched in order from a production process. It is then possible to operate *serial sampling plans* or *deferred sentencing sampling plans,* in which the sentence on a batch depends not only on the results on the batch itself, but also on results from neighbouring or following batches. The plans described below all treat each batch independently; the effect of operating such plans as serial sampling plans would be to modify the sentencing rules depending on the results of inspection on neighbouring batches.

(*i*) *Single sampling plan.* Suppose we have batches of items presented, and the items are to be classified merely as effective or defective. A single sampling plan consists of selecting a fixed random sample of n items from each batch for inspection, and then sentencing each batch depending upon the results. If the sentence is to be either accept or reject the batch, then each batch would be accepted if the number of defectives r found in the n items were less than or equal to the *acceptance number*, c. We summarize as follows:

Single sampling plan:

$$\left.\begin{array}{l}\text{select } n \text{ items,}\\ \text{accept batch if no. of defectives} \leq c,\\ \text{reject batch if no. of defectives} \geq c+1\end{array}\right\} \quad (1.1)$$

For inspection by variables we have a similar sentencing rule. There is no need for the restriction to two terminal decisions and we could have, for example, accept, reject, or sell at a reduced price.

Example 1.5. For the problem of sampling electrical components, Example 1.1, a suitable sampling plan might be to use a single sampling plan with $n = 30$, $c = 2$. □□□

(*ii*) *Double sampling plan.* In this plan a first sample of n_1 items is drawn, as a result of which we may either accept the batch, reject it, or else take a further sample of n_2 items. If the second sample is taken, a decision to accept or reject the batch is taken upon the combined results.

Example 1.6. A double sampling plan for the electrical component sampling problem might be as follows. Select 12 items from the batch and

accept the batch if there are no defectives,
reject the batch if there are 3 or more defectives,
select another sample of 24 items if there are 1 or 2 defectives.

When the second sample is drawn, we count the number of defectives in the combined sample of 36 items and

accept the batch if no. of defectives ≤ 2,
reject the batch if no. of defectives ≥ 3. □□□

A natural extension of double sampling plans is to have multiple sampling plans, with many stages. It is difficult to see how double or multiple sampling plans would be used when there are more than two terminal decisions, unless more than one attribute (or variable) is measured and a much more complex sentencing rule introduced.

(*iii*) *Sequential sampling plan.* A further extension of the multiple sampling idea is the full sequential sampling plan. In this plan, items are drawn from each batch one by one, and after each item a decision is taken as to whether to accept the batch, reject the batch, or sample another item. A simple method of designing sequential sampling plans was discovered by Professors G. A. Barnard and A. Wald during the 1939–45 war. An essential point is that the sample size is not fixed in advance, but it depends on the way the results turn out.

Sequential sampling plans can save a substantial amount of inspection effort, although the overall gain in efficiency is often not great unless inspection is expensive, as is the case in Example 1.2, concerning grading of milk. Another characteristic of plans where sequential sampling can give great gain in efficiency is when the incoming quality is very variable. Again, Example 1.2 provides just such a situation, as the milk being examined comes from many farms over a wide area, and is of very variable quality.

The theory of sequential sampling plans is discussed by Wald (1947) and Wetherill (1975), and will not be discussed further in this book.

(*c*) *Discussion*

We have described many different types of inspection situations and inspection plans, and a number of questions arise. What are the relative merits of different types of plan? How should the sample sizes and acceptance numbers be chosen, and upon what principles? In attempting to answer these questions we should consider carefully the aims for which the inspection plan was instituted. For this reason we discuss the inspection situation in greater detail in the next section. In succeeding chapters we shall discuss the rival theories which have been proposed for the design of sampling plans.

1.4 Flow chart for acceptance inspection

In any realistic assessment of alternative sampling inspection plans, the mechanics of the actual situation into which a sampling plan fits must be considered in some detail. In many papers we find that important – even drastic – assumptions are made, both implicitly and explicitly, as to the manner in which a plan works. In this section we do not attempt to give a complete catalogue of inspection situ-

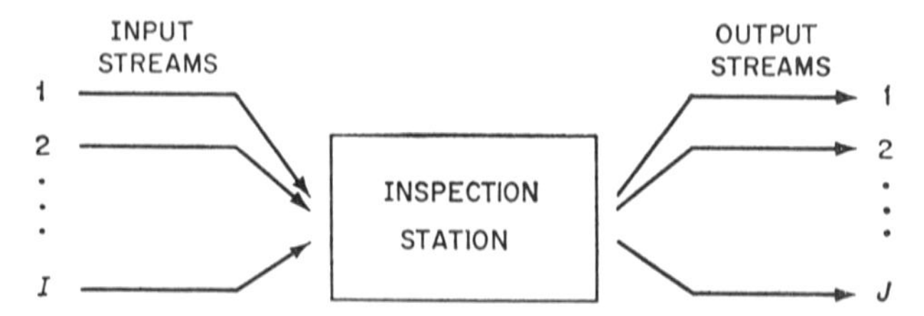

Figure 1.2. *An inspection situation.*

ations, but we aim to give sufficient to form a basis on which to judge the remainder of the book.

Consider the following situation. Batches of approximately N items reach an inspection station through one of I streams. For a consumer, these streams might be different suppliers, while for a producer, the streams might be different production lines; it is possible that the most common case is $I = 1$. The quality of batches in the streams may or may not be correlated with the quality of other neighbouring batches in the same stream or in other streams. It is also possible that these input streams may have different states; for example, a production process may be either in control or out of control. It seems obvious that when several states exist in the input streams, the inspection plan should be specially designed to deal with this.

At the inspection station a sample of items is selected from some or all of the batches and the samples are inspected. Each batch is then sentenced, and placed in one of the J output streams.

If there are only two output streams, these are usually referred to as the *accepted* and the *rejected* batches. For final inspection by a producer, the accepted batches are those passed on for sale to customers. There are many possibilities for the rejected lots, and some of these are set out in Figure 1.3, some of which is taken from Hald (1960). However, this diagram is really appropriate when items are simply classified as effective or defective. More frequently there might be different types of defective, and different action taken on each type.

In some applications of inspection plans there may be more than two output streams. For example, there may be two grades of accepted batches, for different uses, or for sale at different prices. Similarly there could be two grades of rejected batches. However, such plans would often be considered unduly complicated, and liable

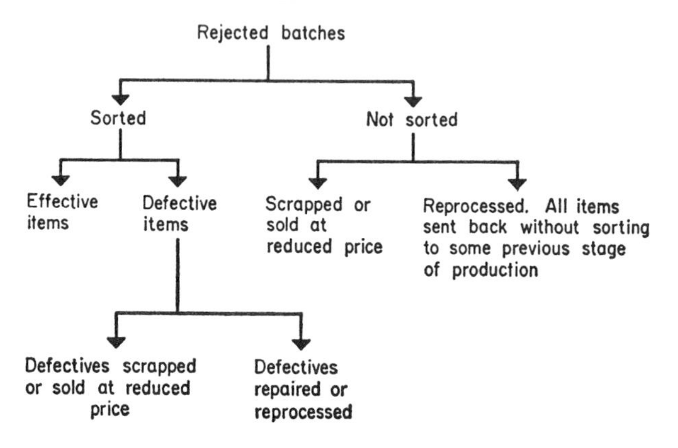

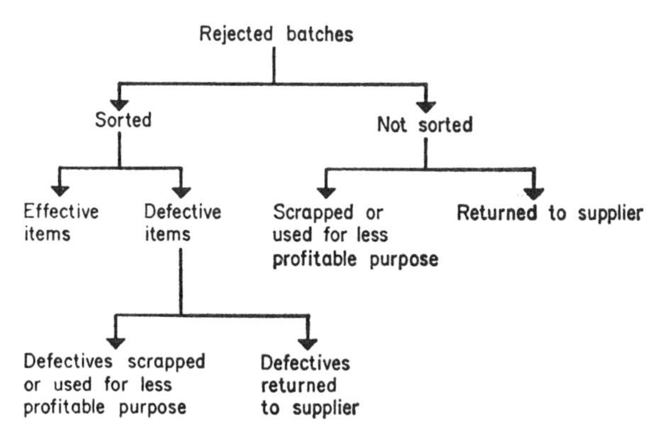

Figure 1.3. *Some possible courses of action on rejected batches.*

to lead to gross errors on the part of the inspector. Here we consider two output streams and call them accepted and rejected batches.

Another point with regard to the flow chart, Figure 1.2, is to specify which parts of this chart work at a given rate, and which parts can work at varying rates. For final inspection by a producer, the input streams are fixed, but for inspection by a consumer, the quantity usually fixed is the number of acceptable batches passed.

In addition to either of these possibilities, the labour and resources available at the inspection station will usually be fixed, and variable only in a long-term sense.

The purpose for which inspection is being applied also needs to be considered in some detail. For a producer, some possible aims are:

(*a*) To satisfy some requirement for the British Standards Institution, etc.
(*b*) To grade batches for sale.
(*c*) To prevent bad batches being passed on to customers.
(*d*) To provide information from which a quality control plan can be operated.

The aims for a consumer might be:

(*e*) To confirm that the quality of goods supplied is up to standard.
(*f*) To prevent bad batches being passed on to a production process.
(*g*) To grade batches for different uses.
(*h*) To encourage the producer to provide the quality desired (Hill, 1960). This purpose can only be achieved if the consumer uses a substantial part of the supplier's output.

It is probable that in many situations in which sampling inspection plans are applied, the aims are not easy to define precisely.

We can see throughout this discussion that inspection by a producer is in general very different from that by a consumer.

An extended discussion of some case studies of quality control practices arising in industry is given by Chiu and Wetherill (1975).

2. Acceptance sampling: basic ideas

2.1 The OC-curve

For the most part this chapter is concerned with inspection situations in which the items are classified as either effective or defective, and where the items are presented in batches. We introduce here some basic concepts and definitions which we shall use in our further discussion; one of the most important of these concepts is the operating characteristic.

Suppose batches of quality θ are presented (so that θ is the proportion which is defective) and the single sample plan (1.1) of section 1.3 is used. That is, n items are selected at random from each batch, and a batch is accepted if c or fewer defectives are found in it.

Then it follows that the probability that a batch of quality θ will be accepted is

$$P(\theta) = \sum_{r=0}^{c} \binom{n}{r} \theta^r (1 - \theta)^{n-r} \tag{2.1}$$

if the finite population corrections are ignored. This function is illustrated in Figure 2.1. Clearly, when $\theta = 0$, all batches are accepted and $P(0) = 1$. As θ increases $P(\theta)$ decreases, until it is zero at $\theta = 1$ (it will be negligible long before $\theta = 1$). This curve shown in Figure 2.1 is called the *operating characteristic curve, or OC-curve.*

For any given sampling plan, the OC-curve can be calculated, and compared with what we think the OC-curve should be like. Ideally, we might wish to have an OC for which all batches with $\theta < \theta'$ were accepted, and all others rejected. This would be

$$P(\theta) = \begin{cases} 1 & \theta < \theta' \\ 0 & \theta > \theta' \end{cases}$$

and is shown in Figure 2.2. This OC-curve is impossible to achieve

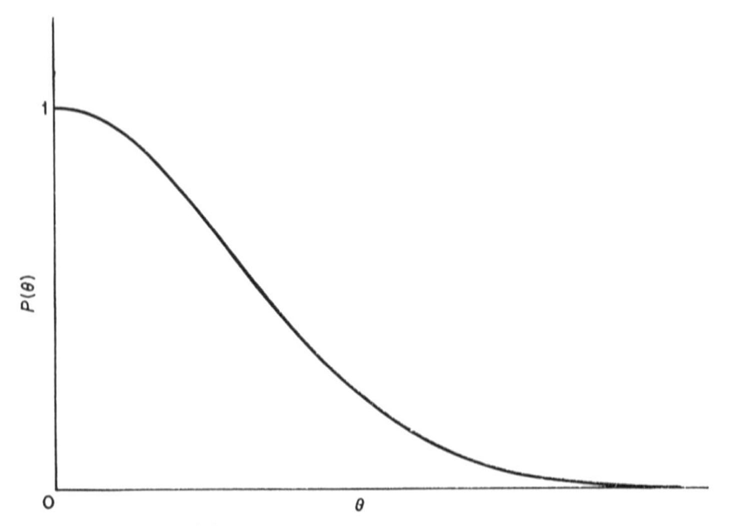

Figure 2.1. *The OC-curve.*

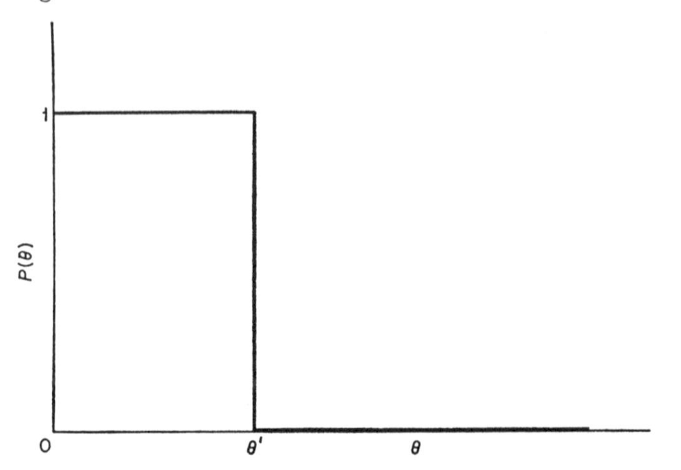

Figure 2.2. *An ideal OC-curve.*

without 100% inspection. An alternative specification would be to set

$$P(\theta) = \begin{cases} 1 & \theta < \theta' \\ 0 & \theta > \theta'' \end{cases} \qquad \theta' < \theta''$$

leaving the region (θ', θ'') in which we do not mind what happens. This is shown in Figure 2.3.

Unfortunately even this alternative form of ideal OC-curve is impossible without 100% inspection, but we can make a specification

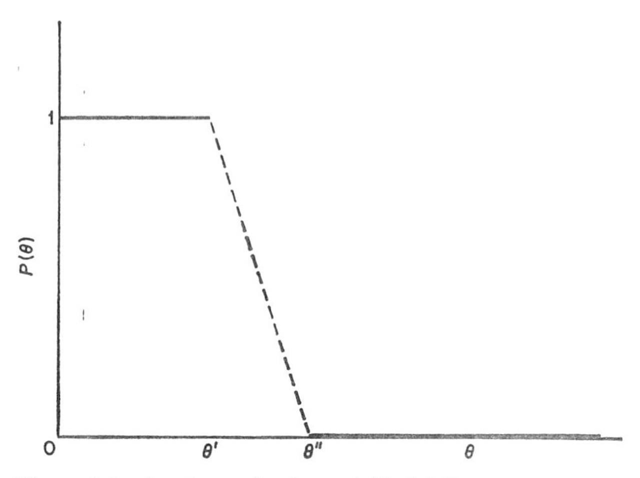

Figure 2.3. *An alternative form of ideal OC-curve.*

which is close to this. Suppose we set $P(\theta)$ to be close to unity at some value θ_1 of θ, say,

$$P(\theta_1) = 1 - \alpha \tag{2.2}$$

and set $P(\theta)$ to be close to zero at some value θ_2,

$$P(\theta_2) = \beta, \qquad \theta_1 < \theta_2 \tag{2.3}$$

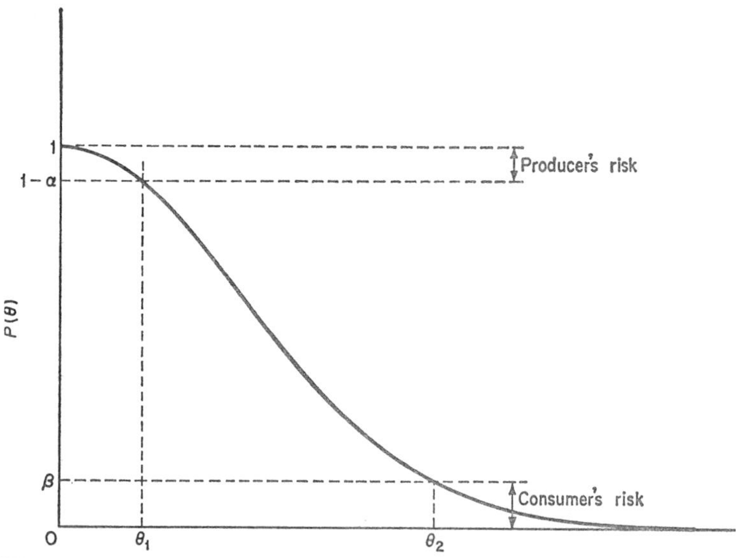

Figure 2.4. *Producer's and consumer's risks.*

then we have a specification we *can* achieve. It means that we are fixing two points on our OC-curve, as shown in Figure 2.4.

The interpretation put upon these points is to postulate rather narrow-minded producers and consumers, as follows. The producer is only concerned that satisfactory batches of quality θ_1 are accepted with a high probability $(1 - \alpha)$. The consumer, on the other hand, is only concerned that batches of unsatisfactory quality, say θ_2, are only accepted with a small probability β. It is realized that no real producer or consumer could define his interests so narrowly.

The points θ_1 and θ_2 are described variously. A suitable terminology is

θ_1 – producers $100(1 - \alpha)$ to 100α safe point
θ_2 – consumers $100(1 - \beta)$ to 100β risk point.

Some ways in which producers' and consumers' risks are used in determining sampling plans are indicated in section 2.4.

Another point on the OC-curve, used by Hamaker and Van Strik (1955), is the central point, or indifference quality or point of control, $\theta_{0\cdot50}$, such that

$$P(\theta_{0\cdot50}) = 0\cdot50. \tag{2.4}$$

Hamaker and Van Strik defined equivalence between sampling plans by the point of control and the slope of the OC-curve at this point, and the authors proceed to compare alternative sampling plans. Clearly, there are a number of other ways of defining equivalence between sampling plans.

Exercises 2.1

1. Calculate and plot the OC-curves for the sampling plans given in Examples 1.5 and 1.6.
2. If we use a single sample plan for per cent defective, and fix the producer's and consumer's risk points, we must solve equations (2.2) and (2.3) for n and c. Suggest a method of solving these equations using the normal approximation to the binomial distribution, and obtain the plan for the parameters $\alpha = \beta = 0\cdot05$, $\theta_1 = 0\cdot01$, $\theta_2 = 0\cdot05$.

When would the normal approximation method be inaccurate?

2.2 The average run length

This concept was introduced by Page (1954); it has since been discussed in many papers, and has found widespread application in industry.

Suppose we have continuous production inspection (of single items), or else batch inspection of an ordered sequence of batches, then the run length is defined as the number of batches (or items) sampled until one is rejected. The distribution of run length for any sampling plan is positively skew with a very long tail. Often we do not consider the whole run length distribution, but limit consideration to the *average run length* or ARL.

Suppose that we have batch inspection using a plan which accepts each batch independently of others with an OC-curve $P(\theta)$. Then the probability that a run length of r batches is observed is

$$\{P(\theta)\}^{r-1}\{1 - P(\theta)\}, \qquad r = 1, 2, \ldots \tag{2.5}$$

and the ARL, is $1/\{1 - P(\theta)\}$. In this situation, therefore, the OC-curve and the ARL function are exactly equivalent. However, it can be argued (particularly in some situations) that the ARL is more directly meaningful. The ARL tells us how much of a given quality is accepted on average, before some action is taken.

In some sampling plans, the plans are altered according to the process average as determined from sampling. For such plans the ARL and OC-curve may not be directly equivalent, and the ARL appears to be the more meaningful concept. Another situation when the ARL should be used is when we have plans which are being used for process control, for the ARL shows how frequently corrective action is initiated. In other cases it may be helpful to use both the ARL and the OC-curve concepts.

Finally, we note that it is sometimes useful to distinguish between the average *sample* run length (ASRL), and the average *article* run length (AARL).

Exercises 2.2

1. Calculate the ARL curve for the sampling plans given in Examples 1.5 and 1.6. (See Exercise 2.1.1.)
2. Calculate some terms of the distribution (2.5) for several values of P, in order to study the shape of the distribution.

2.3 The process curve

The long run distribution of the quality of batches of items arriving at the inspection station is called the process curve. It is possible that a stochastic process of some kind governs the quality of incoming batches, but this is usually ignored in batch inspection, partly on the grounds that it is very difficult in practice to obtain information on the process. With continuing production inspection, there is no meaning to the process curve without either arbitrarily batching it, or else bringing in the stochastic element. A typical process curve is shown in Figure 2.5.

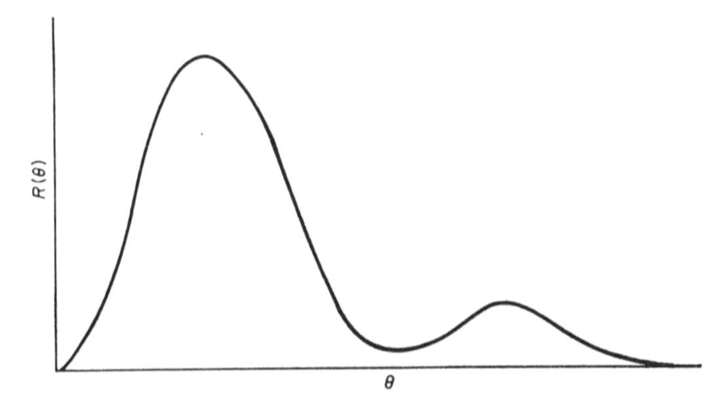

Figure 2.5. *Typical process curve for per cent defective.*

Published data on process curves are very scarce, but some data were collected by Ford (1951), part of which is quoted by Barnard (1954, p. 159). Horsnell (1957) proposes some theoretical models for process curves which he says fit practical data quite well, but Barnard says in the discussion following Horsnell's paper (p. 192) that data he has seen do not bear the slightest resemblance to Horsnell's models. The scarcity of information is due to two reasons:

(*i*) Such data are almost always regarded as industrial secrets.
(*ii*) Production conditions are sometimes not held constant for long enough to accumulate sufficient data.

Hamaker (1958, pp. 151–4) surveys the theoretical models assumed for process curves. The most important are as follows:

$f(\theta) = K\theta^{\alpha}(1 - \theta)^{\beta}$	Champernowne, 1953
$f(\theta) = (s + 1)(1 - \theta)^{s}$	Sittig, 1951
$\Pr(\theta = \theta_i) = a_i, \quad i = 1, 2$	Barnard, 1954
$\Pr(\theta = \theta_i) = a_i, \quad i = 1, 2, \ldots, k$	Generalization of Barnard, 1954
$\Pr(\theta = \theta_i) = a_i, \quad i = 1, 2, \ldots, k,$	where $\theta_i/(1 - \theta_i) = (\theta')^{i/\alpha}$ Wetherill, 1960.

The vital question, of course, is to know how accurately we need a knowledge of the process curve. The research to date indicates that our knowledge of the prior distribution does not need to be very precise (Pfanzagl, 1963; Wetherill, 1960; Wetherill and Campling, 1966) provided that the form of the distribution chosen is reasonable.

It should be pointed out that in nearly all inspection situations *some* knowledge of the process curve is needed to arrive at a satisfactory sampling plan, although this knowledge is often used subjectively; see section 2.4. We need to know (roughly) how likely it is that batches of any given quality will occur, in order to decide how much protection we need at various quality levels.

Mood's theorem

An important result which throws some light on the importance of the process curve was derived by Mood (1943).

Consider a single sample plan for fraction defective, from batches of size N. Any given batch quality can be represented by a point on the batch line in Figure 2.6, and any sample result is represented by a point on the sample line. Consider a batch of quality represented by the point P, then the probability that the sample result is given by Q is

$$\Pr(\mathrm{Q} \mid \mathrm{P}) = \frac{\text{no. of paths OP via Q}}{\text{total no. of paths OP}}$$

$$= \binom{n}{b}\binom{N-n}{B-b} \Big/ \binom{N}{B}$$

Now suppose that the process curve is binomial

$$\binom{N}{B} p^B q^G$$

corresponding to stable production at a probability p of a defective. Then the total probability of obtaining a batch and sample represented by P and Q respectively is

$$\binom{N}{B} p^B q^G \binom{n}{b}\binom{N-n}{B-b} \Big/ \binom{N}{B}$$

$$= \binom{n}{b} p^b q^g \binom{N-n}{B-b} p^{B-b} q^{G-g}.$$

This last statement shows that the sample result (g, b) is *statistically independent* of the quality of the remainder of the batch $(G-g, B-b)$. In particular, b is statistically independent of $(B-b)$.

Mood actually found that the correlation between b and $B - b$ is zero for a binomial process curve, and negative and positive for leptokurtic and platykurtic process curves respectively.

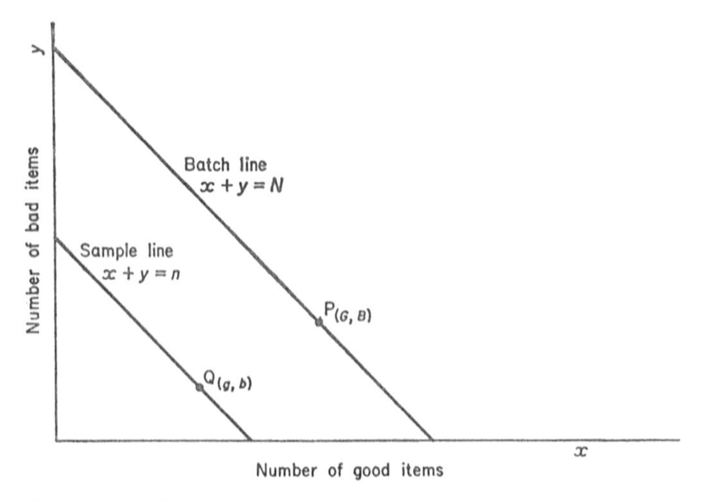

Figure 2.6. *Illustration of Mood's theorem.*

One conclusion which we can draw from this result is that there is no point in sampling when the batch quality is stable (except, maybe, to reject the entire production). Sampling only makes sense with variable quality. We therefore need to take care about schemes worked out on a basis of stable production.

2.4 Methods of choosing sampling plans

The producer's and Consumer's risk point method

In this text we shall use the term sampling scheme to refer to a set of principles used to determine sampling plans, resulting in a collection or table of individual sampling plans indexed ready for use. There are a variety of sampling schemes available, each appropriate in certain circumstances, and we shall discuss the underlying principles of some of these in the next few sections.

Suppose that we have large batches of items presented for acceptance inspection, where the items are classified as effective or defective, then the sampling plan (1.1) has two parameters to be fixed, the sample size n and the acceptance number c; clearly we need two equations to determine these quantities. One way of obtaining two equations is to pick two points on the OC-curve, and determine n and

c so that the OC-curve of our plan goes through (or very near to) these points. Following the reasoning of section 2.1, one convenient pair of points to choose is the producer's risk point (2.2) and the consumer's risk point (2.3); these points are shown in Figure 2.4. Owing to the discreteness of the binomial distribution it may not be possible to satisfy (2.2) and (2.3) exactly, and we can restate them as

$$\sum_{0}^{c} \binom{n}{r} \theta_1^r (1 - \theta_1)^{n-r} \geq 1 - \alpha \tag{2.6}$$

$$\sum_{0}^{c} \binom{n}{r} \theta_2^r (1 - \theta_2)^{n-r} \leq \beta \tag{2.7}$$

We need to find a pair of values (n, c) satisfying these inequalities, and an approximate solution can be obtained as follows. (See Hald (1967) for a further discussion of the method given here, together with approximate solutions etc.)

First we replace the binomial terms in (2.6) and (2.7) by Poisson terms for the same means,

$$\sum_{0}^{c} \mathrm{e}^{-n\theta_1} (n\theta_1)^r / r! \geq 1 - \alpha \tag{2.8}$$

$$\sum_{0}^{c} \mathrm{e}^{-n\theta_2} (n\theta_2)^r / r! \leq \beta \tag{2.9}$$

Now the cumulative Poisson distribution can be related to the cumulative χ^2-distribution since we can show by integration by parts that

$$\frac{1}{c!} \int_{m}^{\infty} t^c \, \mathrm{e}^{-t} \, \mathrm{d}t = \sum_{0}^{c} \mathrm{e}^{-m} m^r / r! \tag{2.10}$$

and hence that

$$\sum_{0}^{c} \mathrm{e}^{-m} m^r / r! = \Pr\{\chi^2 > 2m \mid 2(c + 1) \text{ d.f.}\} \tag{2.11}$$

since the integral on the left-hand side of (2.10) is the probability that χ^2 is greater than $2m$, for a χ^2-distribution having $2(c + 1)$ degrees of freedom. Inequalities (2.8) and (2.9) are therefore equivalent to

$$\Pr\{\chi^2 > 2n\theta_1 \mid 2(c + 1) \text{ d.f.}\} \geq 1 - \alpha \tag{2.12}$$

$$\Pr\{\chi^2 > 2n\theta_2 \mid 2(c + 1) \text{ d.f.}\} \leq \beta \tag{2.13}$$

If we denote the 100α-percentile of the χ^2-distribution with $2(c + 1)$ degrees of freedom by χ^2_α, then these inequalities are

$$2n\theta_1 \leq \chi^2_\alpha \tag{2.14}$$

$$2n\theta_2 \geq \chi^2_{1-\beta} \tag{2.15}$$

If we now put

$$r(c) = \chi^2_{1-\beta}/\chi^2_\alpha \tag{2.16}$$

then c is the smallest value satisfying

$$r(c-1) > \theta_2/\theta_1 > r(c) \tag{2.17}$$

We can solve (2.14) and (2.15) for n to get

$$\chi^2_{1-\beta}/2\theta_2 \leq n \leq \chi^2_\alpha/2\theta_1 \tag{2.18}$$

with the χ^2's having $2(c+1)$ degrees of freedom. Any n in the interval (2.18) will solve the problem, and we can choose which inequality (2.8) or (2.9) is nearer to being satisfied as an equality, by choice of n nearer to one or other limit. If (2.18) does not contain an integral value of n, we must increase c and obtain a new interval (2.18).

In this way a sampling plan approximately satisfying (2.6) and (2.7) is easily obtained, and tables of $r(c)$ and of χ^2 percentage points are given in the Appendix tables. Once an (n, c) is determined, an approximate OC-curve can be plotted using standard χ^2 percentage points and equation (2.11).

Example 2.1. Suppose the parameter values are $(\theta_1 = 0{\cdot}01, \alpha = 0{\cdot}05)$, $(\theta_2 = 0{\cdot}04, \beta = 0{\cdot}05)$, then $\theta_2/\theta_1 = 4$. Looking up Appendix II, table 6, we find that $c = 6$ solves (2.17), giving

$$4{\cdot}02 > 4 > 3{\cdot}60$$

From Appendix II, table 3, we find that the interval for n, (2.18), is

$$23{\cdot}68/0{\cdot}08 \leq n \leq 6{\cdot}57/0{\cdot}02$$

$$296 \leq n \leq 328{\cdot}5$$

Suppose we use $n = 300$, $c = 6$, then the OC-curve is

$$P = \sum_0^6 \mathrm{e}^{-300\theta}(300\theta)^r/r!$$

$$= \Pr\{\chi^2 > 600\theta \mid 14 \text{ d.f.}\}$$

and a few points on this curve can be plotted using Appendix II, table 3. □□□

It is interesting to notice from Example 2.1 that $c = 5$ will nearly satisfy the requirements. The appropriate n interval is then

$$261{\cdot}3 \geq n \geq 262{\cdot}8$$

which cannot be satisfied. If we use, say ($n = 262$, $c = 5$), then the actual risks will be slightly greater than those set. In view of the great reduction of sample size, we would probably use this plan in this example.

We notice that the producer's and consumer's risk points specified in Example 2.1 have led to an extremely large sample size. If this were a practical case, we would seek to modify the risk points and try again etc., until we arrived at a sampling plan which is felt to be 'reasonable'. This type of iterative process can be defended on the grounds that one is trying to balance the cost of sampling against the costs of wrong decisions. For any batch of quality θ, the probability that it will be accepted is given by the OC-curve. Thus by looking at the OC-curve, we can see the probability that poor quality will be accepted and good quality rejected. The probabilities of these wrong decisions can be reduced – but only by increasing the sample size and so increasing sampling costs. The essential point about the balancing of costs referred to here is that it is not formalized. The final decision on a sampling plan is made subjectively, by someone with a detailed knowledge of the set-up.

A number of tables of sampling plans have been constructed based upon principles rather similar to the above. Peach (1947) listed sampling plans for which the producer's and consumer's risks were both set at 0·05. Horsnell (1954) tabulated plans for producer's risks of 0·01 or 0·05 and consumer's risks of 0·01, 0·05, or 0·10.

The most comprehensive set of tables of this kind is provided by Hald and Kousgaard (1966). Essentially they give tables of

$$\sum_{n=0}^{c} \binom{n}{r} \theta^r (1 - \theta)^{n-r} = P$$

for $c = 0(1)\,100$, 15 values of P, and values of $0 < \theta < 0{\cdot}50$. Simple illustrations are given of the use of these tables to obtain a single sampling plan with set producer's and consumer's risks.

Two main criticisms can be levelled at the producer's and consumer's risks method of determining sampling plans. The first is that, except in very small batch sizes, the resulting plans are independent of the batch size. (Clearly in very small batch sizes the hypergeometric distribution should be used in (2.6) and (2.7).) Since the costs of wrong decisions increase with batch size, it is obvious that the probabilities of error (α, β) should reduce with increasing batch size.

The second criticism is that it is in general rather difficult to choose the parameters (θ_1, α; θ_2, β). If we are dealing with an endless sequence of batches, the OC-curve points could be expressed as ARL's, which might have more meaning, but the choice has to be made in consultation with production staff and others who do not appreciate the full depth of the concepts involved.

It is important to realize that the OC-curve does *not* give the proportion of batches of any given quality among accepted batches, since it is necessary to use the process curve to obtain this quantity. The probability distribution of θ among accepted batches is clearly

$$P(\theta).R(\theta) \Big/ \int_0^1 P(\theta)\, R(\theta)\, \mathrm{d}\theta \tag{2.19}$$

where $P(\theta)$ is the OC-curve and $R(\theta)$ the process curve. Thus the proportion of accepted batches having $\theta > \theta_1$ is

$$\int_{\theta_1}^1 P(\theta)\, R(\theta)\, \mathrm{d}\theta \Big/ \int_0^1 P(\theta)\, R(\theta)\, \mathrm{d}\theta \tag{2.20}$$

The effect of a sampling plan is to change the distribution of batch quality from the distribution $R(\theta)$ which is input, to a distribution proportional to $P(\theta)\, R(\theta)$ which is output from the inspection station. When using an OC-curve sampling scheme we have to keep this in mind. Figure 2.7 illustrates the effect of a sampling plan.

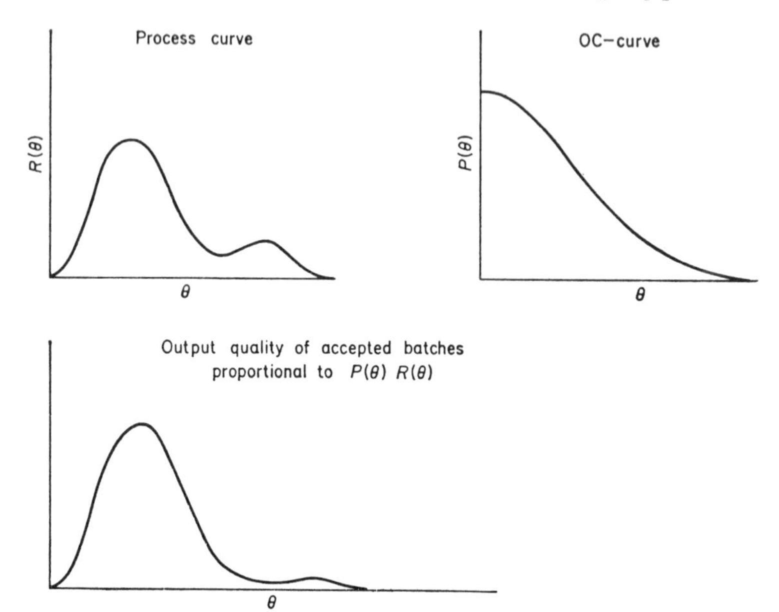

Figure 2.7. *Effect of a sampling inspection plan.*

Exercises 2.4

1. Compare the accuracy of the method given in section 2.4 for choosing a single sample plan with the normal approximation method used in Exercise 2.1.2.
2. Plot and compare the OC-curves for the two plans suggested for Exercise 2.1, ($n = 300$, $c = 6$) and ($n = 262$, $c = 5$).
3. Use the Poisson approximation to design a single sample plan for the producer's and consumer's risk method with:

(*a*) $\theta_1 = 0{\cdot}01$, $\alpha = 5\%$; $\theta_2 = 0{\cdot}05$, $\beta = 10\%$
(*b*) $\theta_1 = 0{\cdot}01$, $\alpha = 5\%$; $\theta_2 = 0{\cdot}04$, $\beta = 10\%$
(*c*) $\theta_1 = 0{\cdot}01$, $\alpha = 5\%$; $\theta_2 = 0{\cdot}03$, $\beta = 10\%$

Finally, write a short note on the effect of changes in the ratio θ_2/θ_1 on sample size. (Use Appendix II table 6.)
4. What is the effect on the calculations given in question 3 of

(*a*) reducing β to $2{\cdot}5\%$?
(*b*) multiplying θ_1 and θ_2 by 2?
(*c*) reducing α to $2{\cdot}5\%$?

2.5 Defence sampling schemes

A series of sampling schemes have been developed during and since the Second World War, for use in military contracts, and they are all based on similar principles.

The first step is that a table was drawn up which in effect fixes the relationship between batch size and sample size to be one of three or five purely arbitrary functions. The sample sizes were made to increase with batch size in a manner thought to be reasonable.

Next the concept of Acceptable Quality Level (AQL) was introduced, but the actual definition of this differs in the different schemes. The SRG tables (Freeman *et al.*, 1948) fixed the AQL as the quality for which the probability of acceptance was 0·95. Unfortunately this has some undesirable consequences: since the sample size is already fixed, this automatically determines the sampling plan, and some rather large consumer's risks result. Other sampling schemes, such as the U.S. Army Service Forces tables (1944), MIL-STD-105 (A, B, C, and D), and the British DEF-131 (Hill, 1962), have let the probability of acceptance at the AQL vary in a rather unsystematic way, so as to share the risks between producer and consumer more equitably. The most satisfactory definition of AQL is the one used in DEF-131, that it is the maximum percentage defective which can be considered satisfactory as a process average. That is, the AQL is a property required of the product. The variation of the probability of acceptance at AQL is considerable, ranging from 0·80 for the lowest sample sizes to 0·99 for the largest.

Finally, all of the defence sampling tables use *switching rules*. The idea is that watch is kept on the inspection results, and according to certain rules, 'tightened' inspection is introduced if necessary. The introduction of tightened inspection produces a wholesale change of the OC-curve along the lines shown in Figure 2.8. This puts considerable pressure on the producer, since even goods of AQL quality would be rejected much more frequently. It should be clear that a major part of the quality assurance given by the schemes lies in the use of this switching rule pressure tactic. A producer is forced to send goods of AQL quality or better, to have them accepted at a satisfactory rate.

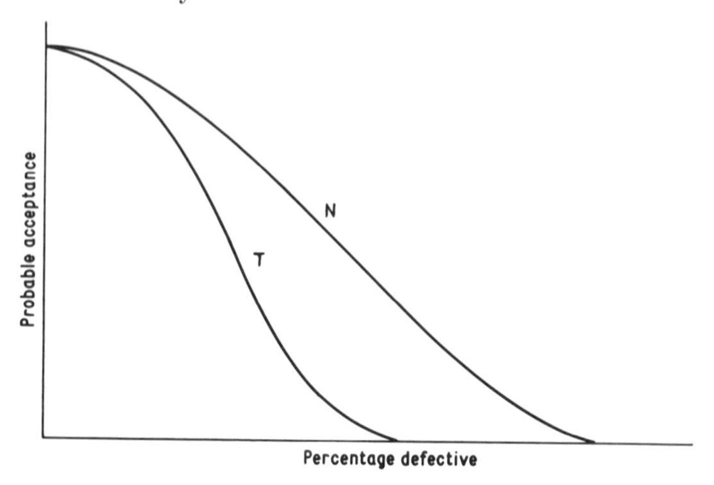

Figure 2.8. *OC-curves for normal* (N) *and tightened* (T) *inspection.*

The rules used for switching have varied. Ideally, a watch should be kept on the process average, but this was criticized as being too complicated. A frequently used rule is to switch to tightened inspection if two or more of the last five batches have been rejected.

A good description of a modern scheme, together with some tables, is readily accessible in the paper by Hill (1962), to which readers are referred for further details. A complete description of the schemes will not be given here.

There are two main criticisms of the Defence Sampling Schemes method. One obvious point is the arbitrariness of the batch-sample size relationship. There is no theoretical backing for any of the relationships used. A rather more important criticism, however, is that the switching rules pressure tactic is not always practicable. There may not be an indefinite sequence of batches, and the consumer may not be in a position to exert much pressure on the producer. (A government department is usually in a different position.) However, in places where switching rules can be used, the Defence Sampling Schemes method probably pays off well.

Particular schemes of this type which are widely used are DEF-131A and MIL-STD-105D.

2.6 Decision theory schemes

An alternative scheme for designing a sampling plan is to assess the costs and losses involved and try to design the plan so as to minimize costs. Increased sampling (which costs more) will reduce losses from wrong decisions; hence there will be an economic optimum which can be determined.

Consider Example 1.1, the problem concerning electronic components. Let the cost of sampling an item be regarded as the unit of costs, denote the profit from passing a good item by b, and denote the profit (loss) from passing a bad item by $-f$. In Example 1.1 this latter quantity is the cost of tracing and eliminating faulty components. The profit from a batch of size N and quality θ is then

$$N(1 - \theta)b - N\theta f = k(\theta_0 - \theta)$$

say, where $k = N(b + f)$, and $\theta_0 = b/(b + f)$ is the *break-even quality* at which it is just as profitable to accept or reject a batch. Clearly, we wish to accept batches for $\theta < \theta_0$ and reject batches having $\theta > \theta_0$, and the decision losses of wrong decisions can be tabulated as follows:

Decision	*Quality*	*Loss*
Accept	$\theta > \theta_0$	$k(\theta - \theta_0)$
Accept	$\theta < \theta_0$	0
Reject	$\theta > \theta_0$	0
Reject	$\theta < \theta_0$	$k(\theta_0 - \theta)$

Suppose the prior distribution of the quality of incoming batches for inspection is beta,

$$\theta^{\alpha-1}(1 - \theta)^{\beta-1}\Gamma(\alpha + \beta)/\Gamma(\beta).\Gamma(\alpha), \quad 0 < \theta < 1$$

We shall now show how to determine the values of n and c which describe the simple sampling plan for which the expected loss is a minimum. As a first step we find the value of c which is optimum for a given n.

For any n and r the posterior loss of rejection is

$$\int_0^{\theta_0} k \mid \theta - \theta_0 \mid \theta^{X-1}(1-\theta)^{N-1}\,\mathrm{d}\theta/\beta(X, N) \tag{2.21}$$

where $X = \alpha + r$ and $N = n + \beta - r$. Similarly, the loss of acceptance is

$$\int_{\theta_0}^{1} k \mid \theta - \theta_0 \mid \theta^{X-1}(1-\theta)^{N-1}\,\mathrm{d}\theta/\beta(X, N). \tag{2.22}$$

To minimize losses, we accept or reject at any point (n, r) according to which of (2.21) and (2.22) is the smaller. The acceptance number c will be where the two are equal. This leads to the equation

$$c = (n + \alpha + \beta)\theta_0 - \alpha - \tfrac{1}{2}. \tag{2.23}$$

As we might expect, $c \simeq n\theta_0$.

The determination of the optimum sample size is more complicated. We evaluate the total losses and minimize them for choice of n. For a process curve $\mathrm{R}(\theta)$ the losses are

$$\begin{aligned} R &= (\text{cost of sampling}) + (\text{cost of wrong decisions}) \\ &= n + \int_{\theta_0}^{1} k \mid \theta - \theta_0 \mid P(\theta)R(\theta)\,\mathrm{d}\theta \\ &\quad + \int_0^{\theta_0} k \mid \theta - \theta_0 \mid \{1 - P(\theta)\}R(\theta)\,\mathrm{d}\theta \\ &= n + \int_0^{1} k(\theta - \theta_0)P(\theta)\,\mathrm{d}\theta + \int_0^{\theta_0} k(\theta_0 - \theta)R(\theta)\,\mathrm{d}\theta. \end{aligned} \tag{2.24}$$

The last term in (2.24) is independent of n, and the relationship (2.23) should be used to substitute for c in $P(\theta)$, in the middle term. If equation (2.24) is minimized for choice of n we have the optimum sample size.

One useful prior distribution for which the mathematics turns out to be reasonably simple was introduced by Barnard (1954). This is the two-point binomial prior distribution, where θ takes on one of two values, θ_1 or θ_2, with

$$\Pr(\theta = \theta_1) = a, \quad \Pr(\theta = \theta_2) = 1 - a, \quad 0 < a < 1. \tag{2.25}$$

Single sampling plans for the two-point binomial prior distribution were tabulated by Hald (1965). Hald also gives asymptotic and approximate formulae for the optimum plans. A simple method of solution is given by Wetherill (1960).

When using the decision-theory scheme just described, the parameters of the prior distribution and of the loss functions will have to be estimated. A vital question is therefore to examine how robust the economic optimum is to errors in these estimated parameters, and this problem was examined by Hald (1965), and Wetherill and Campling (1966). The general conclusion is that the optimum is very robust to all parameters except the break-even quality θ_0 and the surface is very flat near the optimum for choice of sample size. In general, asymptotic formulae can be used instead of exact formulae with very little loss, see Hald (1967).

The main criticisms of the decision-theory schemes concern the difficulty of estimating the parameters of the loss functions and of the prior distribution. Sometimes production is done in short bursts, so that one type of product is made for a week or two, then another, and so on. In this situation there may be no knowledge at all of the prior distribution. Also, Exercise 1.1 studied above was such that the profits and losses could be determined readily. In situations where part of the cost of passing bad items is loss of goodwill, a more difficult problem arises in determining the losses. The robustness studies referred to in the paragraph above are a partial answer to these criticisms, provided the break-even quality is determined precisely.

Another argument against decision theory schemes is that they assume a constant process curve whereas in some cases one aim of inspection is to force the producer to deliver better quality, and hence to change the process curve. However, the robustness studies have shown that decision theory schemes are not very sensitive to the process curve used. In any case, the real question is whether a better sampling plan is likely to emerge from a decision theory approach than another approach, and not whether the actual optimum plan is achieved. Undoubtedly, there are many situations where a decision theory approach should be used.

2.7 A simple semi-economic scheme

Wetherill and Chiu (1974) have followed up the decision theory work by proposing a simple but highly efficient scheme.

The decision theory work shows that the break-even quality θ_0 is by far the most important parameter, and that under a very wide range of conditions we have $c \simeq n\theta_0$, or to a better approximation

$$c \simeq n\theta_0 - 2/3 \tag{2.26}$$

If we obtain an estimate of the break-even quality, this equation can be used as the starting point of a sampling scheme. We need one more restriction, and the decision theory robustness studies showed that it does not particularly matter, within a wide range, concerning the determination of sample size.

It seems intuitively obvious that we ought to concentrate on properties of the plan with respect to the quality most frequently offered, or the quality required, which we shall call the AQL here. It follows from some work of Hald (1964) that if we make

$$\Pr(\text{accept batch at AQL}) = 1 - \alpha/N \qquad (2.27)$$

for some α, where N is the batch size, then the resulting relationship between n and N is approximately the same as those obtained theoretically for a range of process curves.

In this scheme therefore we determine θ_0 from economic considerations, and then use (2.26) and (2.27) to determine a sampling plan. Wetherill and Chiu (1974) produced a table to simplify this. The only arbitrary choices here are the AQL and α in (2.27), but this enables one to set the probability of acceptance at the quality desired.

Clearly, under this scheme a producer will have to offer quality of about AQL, but it could be slightly worse and still be accepted with high probability.

Exercises 2.7

1. By using equation (2.26) and the Poisson approximation formula (2.11), find the single sample plan with break-even quality approximately $\theta_0 = 0{\cdot}03$, and a producer's risk of 5% at $\theta = 0{\cdot}01$. Plot the OC-curve of your sampling plan.
2. What is your solution to question 1 if either (*a*) $\theta_0 = 0{\cdot}05$, or (*b*) $\alpha = 2{\cdot}5\%$?

2.8 Dodge and Romig's schemes

Dodge and Romig (1929, 1959) pioneered sampling inspection from about 1920 onwards, and they proposed two different schemes, both assuming that rectifying inspection is involved. That is, they assume that rejected batches are 100% sorted, and all defective items replaced or rectified.

One approach was through a quantity they defined as the Lot Tolerance Percent Defective, LTPD, which is 'some chosen limiting value of per cent defective in a lot', representing what the consumer regards as borderline quality. The tables always used LTPD with a consumer's risk of 0·10, so that the LTPD is effectively the quality corresponding to a consumer's risk of 0·10. However, the LTPD only gives us one restriction, and two are required to determine a single sample plan. For the second restriction, Dodge and Romig minimized the average amount of inspection at the process average. For sample and batch sizes n and N, and acceptance number c, the average amount of inspection is

$$I = n + (N - n)\left[1 - \sum_{0}^{c} \binom{n}{r} \theta^r (1 - \theta)^{n-r}\right] \tag{2.28}$$

Subject to the LTPD, values of (n, c) were chosen to minimize I at the process average quality, $\bar{\theta}$.

The other approach Dodge and Romig used involved a quantity defined as the Average Outgoing Quality Limit, AOQL. To obtain this we notice that if on average I items per batch are rectified, an average of $(N - I)$ remain unrectified. The average outgoing quality is therefore

$$\text{AOQ} = (N - I)\theta/N \tag{2.29}$$

where I is given by (2.28). It is readily seen that the AOQ has a graph roughly as shown in Figure 2.9 and passes through a maximum with respect to θ, and there is an upper limit to the average outgoing percent defective called the AOQL.

In the second approach, Dodge and Romig produce sampling

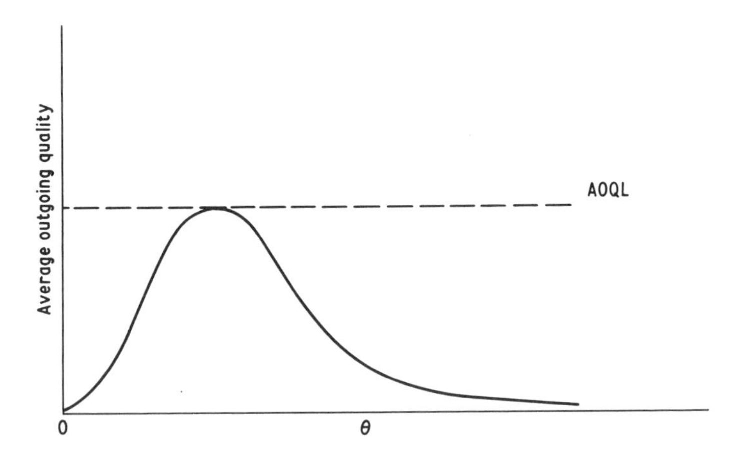

Figure 2.9. *The average outgoing quality limit.*

plans having set values of the AOQL, which also minimize the average amount of inspection at the process average. If all rejected batches are 100% inspected and an average quality guarantee is satisfactory, the AOQL approach may be a good scheme to use.

Some criticisms of Dodge and Romig's methods are given by Hill (1962). In particular he criticizes the AOQL concept as being very sensitive to imperfect inspection, as there then may be no upper limit to the AOQ.

2.9 Other schemes

There are other schemes for deriving sampling inspection plans, and we mention two very briefly.

Horsnell (1957) has an interesting formulation which is a compromise between methods of sections 2.4 and 2.6 above, and which avoids estimating certain costs and losses. He considers inspection by a consumer, and fixes the consumer's risk point. Rejected lots are reckoned (in the long run) to be a total loss to the consumer, and Horsnell minimizes the costs of inspection and the costs of rejected lots, per lot accepted, subject to the fixed consumer's risk point.

A rather similar approach has been adopted by Hald (1964); decision losses are minimized for a two-point binomial prior distribution (2.25) subject to *one* of the following three points on the OC-curve being fixed:

(*i*) Producer's 5% risk point fixed at θ_1 (AQL scheme)
(*ii*) Consumer's 10% risk point fixed at θ_2 (LTPD scheme)
(*iii*) The indifference quality level is fixed at $\theta_{0.50}$ (IQL scheme)

The sample size can then be chosen to minimize losses, and tables are provided to do this. The IQL scheme is a particularly attractive one, and Hald (1965) shows that it has a number of very desirable properties. The indifference quality level should be chosen equal to the break-even quality.

Hill (1960) put forward an entirely different type of economic scheme. One purpose of an inspection plan is taken to be that of inducing a manufacturer to change his prior distribution. Improved quality costs more, both to manufacturer and consumer. If the manufacturer's quality deterioriates, the cost of rejected batches becomes prohibitive. A sampling plan is therefore devised such that it is to the manufacturer's economic advantage to offer the quality desired by the consumer. This scheme is called the *economic incentive* scheme; it implies that the consumer's purchases from the manufacturer are so large that he is willing to make alterations to his process specially for him. This may be true for government purchases and certain large industries, but it clearly does not apply generally.

In the discussion given above, it has been assumed that each batch is to be sentenced independently of other batches, the only break to this rule being the application of 'switching rules' in Defence Sampling Schemes. However, some plans have been proposed which use the results of inspection on neighbouring batches when sentencing any given batch. These plans can only be used if certain conditions are fulfilled, that the batches are part of a continuing supply, that the customer has confidence in the supplier not to insert a bad batch in a sequence, and that there is no reason to believe the batch under consideration is of worse quality than the neighbouring batches. One plan of this sort was a 'chain sampling plan' devised by Dodge (1955). This plan applies where a very small sample size is used, with acceptance number $c = 0$. The idea is simply that a batch having one defective in the sample is accepted if the preceding i samples are free of defects.

A more general application of this idea was discussed by Cox (1960), who called his plans 'serial sampling plans'. He assumed that the process curve was the result of a two-state Markov chain, and he showed how decision theory arguments can then be used to obtain an acceptance rule which involves the results of sampling on any specified neighbouring batches.

2.10 Discussion

There is an arbitrary element in any scheme for choosing a sampling plan, although this is very much less marked for decision theory approaches. In OC-curve schemes everything depends on a suitable choice for producer's and consumer's risk points, and this is more difficult than one might think, except perhaps for experienced users in familiar situations.

Decision theory schemes are more precise and scientific, leaving much less to judgement. However, the process curve and loss functions will have to be chosen, and it is sometimes difficult to obtain information about these quantities. Frequently one of the aims of a sampling plan is to force modifications to the process curve, whereas decision theory plans assume this to be stationary, in a long-term sense. These arguments against decision theory schemes lose much of their force when we see how robust the optimum loss is to errors in the assumed process curve and loss functions. For the example we have discussed, the break-even quality θ_0 is the only critical parameter, and any plan with an acceptance number very different from $c = n\theta_0$ is inefficient. Since OC-curve schemes do not attempt to estimate θ_0 it is clear that large errors are possible. This work indicates therefore that an attempt should always be made to estimate the break-even quality, even in situations where this might be difficult to do.

In factories where production is carried out in short runs of different kinds of product, information on the process curve would be virtually impossible to obtain. In this situation a suitable compromise would be as follows. The break-even quality θ_0 is estimated and then the acceptance number put at $c = n\theta_0$. The sample size n can then be chosen so that the plan has suitable OC-curve properties. With this compromise the difficulty of estimating the process curve can be postponed until more information is available.

Another difficulty with OC-curve schemes, which is also possessed by the compromise scheme suggested above, is that the sample size is not dependent on the batch size. In decision theory schemes an increased batch size leads to increased losses, and thus to increased sample size, and this seems intuitively correct. In some OC-curve approaches this point is dealt with by using an arbitrary relation between batch size and sample size, and although this seems more correct it cannot be said to be totally satisfactory.

Many of the sampling schemes referred to above include double and multiple sampling plans, but the discussion has been based on single sampling plans so as to concentrate on the underlying principles. The usual way to obtain double sampling schemes is to apply some arbitrary restrictions so as to reduce the number of parameters involved. In some cases, these arbitrary restrictions have not been applied in a very sensible way, see Hill (1962), and Hamaker and Van Strik (1955). Sequential sampling plans have always followed use of the sequential probability ratio test; see Wetherill (1975).

There is no problem in deriving double and sequential sampling plans by the decision theory approach, and for details see Wetherill (1975) and Wetherill and Campling (1966).

The final choice of a sampling scheme must be the one which we think is likely to lead to a more efficient sampling plan. In some cases it may be better to try two or three schemes before deciding on a particular plan.

Exercises 2.10

1. Write an essay summarizing the advantages and disadvantages of the methods of designing attributes sampling inspection plans which are discussed in this chapter.

2.11 Inspection by variables

(a) *General*

So far we have been discussing inspection by attributes, when inspected items are simply classified as effective or defective, according to whether they do or do not possess certain attributes. When a continuous variable can be measured upon which the quality or acceptance of an item depends, it may be possible to operate an 'inspection by variables' plan. This will usually lead to a substantial saving of effort over an 'inspection by attributes' plan.

A good discussion of inspection by variables with theory, tables and references, is given by Lieberman and Resnikoff (1955), and their tables are reproduced in the U.S. Military Standard 414. A more detailed coverage of inspection by variables, including tables, worked examples and problems for solution, is given by Duncan (1974). In this section we give only a brief discussion of the main principles.

Consider the following examples, which are quoted by Lieberman and Resnikoff.

Example 2.2. A batch of steel castings is presented for acceptance inspection, and any casting with a yield point below 55,000 lb. per square inch (p.s.i.) is unacceptable. Six castings were tested and the yield points were (in 10^3 p.s.i.)

62·0, 61·0, 68·5, 59·5, 65·5, 63·9.

If certain assumptions are made about the distribution of yield points of castings in the batch, the percentage of defectives can be estimated and the batch sentenced. □□□

Example 2.3. A specification for certain electrical components states that the resistances must be between 620 and 660 ohms. Measurements on the resistance of a random sample of ten items from a large batch give the following results:

639, 640, 650, 647, 662, 637, 652, 643, 657, 649.

□□□

In Example 2.2 there is a single (lower) specification limit for deciding when an item is acceptable, whereas in Example 2.3 there is a double specification limit. We shall discuss Example 2.2 first.

If there are very large numbers of castings in the batches referred to in Example 2.2, we may be able to assume, say, that the distribution of yield points is approximately normal in each batch. Frequently we find that in such a case variations in quality arise by variations in the mean, but the standard deviation σ is nearly constant from batch to batch. If we assume σ to be constant, we have two cases:

(*i*) σ known from past data, and
(*ii*) σ unknown.

When σ is unknown we may use either the sample standard deviation

$$s = \sqrt{\left\{\sum_{1}^{r} (x_i - \bar{x})^2/(n-1)\right\}},$$

to estimate σ, or else we may use the sample range

R_n = range = (largest observation) − (smallest observation).

It can be shown that

$$E(R_n) = \alpha_n \sigma,$$

so that an estimate of σ can be obtained by dividing R_n by α_n. A short table for conversion of ranges to estimates of σ is given in Appendix II, table 4; the table gives $a_n = 1/\alpha_n$.

(*b*) *Single specification limit, σ known*

Let us consider the simplest situation, and take Example 2.2, assuming the distribution of yield points in a batch to be normal with a $\sigma = 3{\cdot}0 \times 10^3$ p.s.i. known from past data. The average of the six results in Example 2.2 in 63·4, so that using this as an estimate of the mean of the batch, the situation is as illustrated in Figure 2.10.

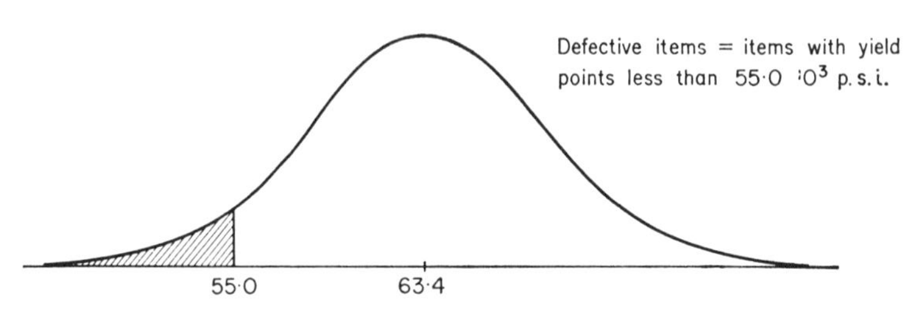

Figure 2.10. *Estimation of percentage defective for Example 2.2.*

We could estimate the percentage of defective items as the proportion of the distribution less than 55·0; this is

$$\Phi\left(\frac{55{\cdot}0 - 63{\cdot}4}{3{\cdot}0}\right) = 1 - \Phi(2{\cdot}8) = 0{\cdot}0026. \tag{2.30}$$

Lieberman and Resnikoff showed that a better estimate of the percentage defectives is obtained by using the formula

$$\tilde{p} = 1 - \Phi\left\{\left(\frac{\bar{x} - L}{\sigma}\right)\sqrt{\left(\frac{n}{n-1}\right)}\right\}, \tag{2.31}$$

where $\bar{x}$ is the mean of the observed sample of n, and L is the lower specification limit. (Parallel formulae apply for an upper specification limit and double specification limits.) For Example 2.2, formula (2.31) gives

$$\tilde{p} = 1 - \Phi\left\{\frac{63{\cdot}4 - 55{\cdot}0}{3}\sqrt{\left(\frac{6}{5}\right)}\right\} = 0{\cdot}0011.$$

Having obtained an estimate of the percentage defective in the batch, we can now accept or reject the batch according to whether $\tilde{p} < p^*$ or $\tilde{p} > p^*$, where p^* is some chosen number.

The sampling plan just described for Example 2.2 involves two parameters, the sample size n, and the acceptance percentage p^*. These two parameters can be chosen according to any of the principles outlined in previous sections. One possibility is to use an OC-curve sampling scheme in which the sampling plan is required to accept a proportion $(1 - \alpha)$ of batches which contain $100\theta_1\%$ defectives, and accept a proportion β of batches which contain $100\theta_2\%$ defectives, where (θ_1, α), (θ_2, β) are given. This can be done easily, as follows.

The decision rule is equivalent to accepting when

$$\left(\frac{\bar{x} - L}{\sigma}\right)\sqrt{\left(\frac{n}{n-1}\right)} > Z_{p^*}$$

where
$$p^* = \int_{Z_{p^*}}^{\infty} \frac{1}{\sqrt{(2\pi)}} \mathrm{e}^{-\frac{1}{2}t^2}\, dt.$$

This is equivalent to accepting when

$$\bar{x} > L + \sqrt{\left(\frac{n-1}{n}\right)} Z_{p^*}\sigma, \tag{2.32}$$

and $\bar{x}$ is normally distributed $E(\bar{x}) = \mu$, $V(\bar{x}) = \sigma^2/n$.

Now for a batch to contain $100\theta_1$ % defectives when σ is known, the mean must be at

$$\mu = L + Z_{\theta_1}\sigma$$
$$= \mu_1 \text{ (say)},$$

and for a batch to contain $100\theta_2$ % defectives the mean must be at

$$\mu = L + Z_{\theta_2}\sigma$$
$$= \mu_2 \text{ (say)}.$$

We require the probability that (2.32) is true to be $(1 - \alpha)$ when $\mu = \mu_1$, and β when $\mu = \mu_2$. Now (2.32) can be written

$$\frac{(\bar{x} - \mu)}{\sigma/\sqrt{n}} > \frac{(L - \mu)}{\sigma}\sqrt{n} + \sqrt{(n-1)}\, Z_{p^*} \tag{2.33}$$

and the quantity on the left-hand side of (2.33) is a standard normal variable. The probability of accepting is therefore

$$1 - \Phi\left\{\frac{(L - \mu)}{\sigma}\sqrt{n} + \sqrt{(n-1)}\, Z_{p^*}\right\}$$

which must be set at $(1 - \alpha)$ when $\mu = \mu_1$, and β when $\mu = \mu_2$. We therefore have to solve the following equations for n and Z_{p^*},

$$-Z_{\theta_1}\sqrt{n} + \sqrt{(n-1)}\, Z_{p^*} = -Z_\alpha$$
$$-Z_{\theta_2}\sqrt{n} + \sqrt{(n-1)}\, Z_{p^*} = +Z_\beta$$

which gives
$$n = \left(\frac{Z_\alpha + Z_\beta}{Z_{\theta_1} - Z_{\theta_2}}\right)^2 \tag{2.34}$$

and
$$\sqrt{\left(\frac{n-1}{n}\right)}\, Z_{p^*} = Z_{\theta_1} - \frac{1}{\sqrt{n}} Z_\alpha, \tag{2.35}$$

or
$$\sqrt{\left(\frac{n-1}{n}\right)}\, Z_{p^*} = Z_{\theta_2} + \frac{1}{\sqrt{n}} Z_\beta. \tag{2.36}$$

The quantity on the left-hand side of (2.35) and (2.36) is the quantity used in the decision rule (2.32). Because of discreteness difficulties with n, the solutions of (2.35) and (2.36) will not be identical, and the average should be used.

The OC-curve of the procedure is easy to derive. If there are $100p$ % defectives in the batch the mean must be at

$$\mu = L + Z_p\sigma.$$

The probability of acceptance is the probability that (2.32) is true, which is

$$\Pr\left\{\bar{x} > L + \sqrt{\left(\frac{n-1}{n}\right)} Z_{p^*}\sigma \,\middle|\, \bar{x}\, N(\mu, \sigma^2/n)\right\}$$
$$= \Phi\{\sqrt{n}\, Z_p - \sqrt{(n-1)}\, Z_{p^*}\}. \quad (2.37)$$

This is the OC-curve of this procedure.

(c) *Double specification limit, σ known*
We now consider briefly the case illustrated in Example 2.3, where there is a two-sided specification, (L, U) for acceptable components.

If $(U - L)/\sigma$ is large, say greater than 6, a sampling plan can be derived by merely combining two one-sided plans, and all the formulae will be approximately correct. However, there is a minimum

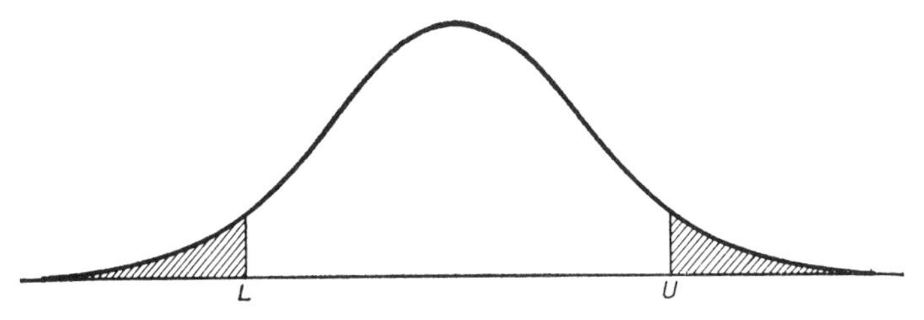

Figure 2.11. *Double specification limit.*

percentage of defectives, corresponding to a mean in the centre of the range (L, U); the percentage defectives is then

$$2\Phi\{(L - U)/2\sigma\} \quad (2.38)$$

and if $(U - L)/\sigma$ is small enough, the batch can be rejected outright. Hence it is intermediate values of $(U - L)/\sigma$ which cause difficulty, and the situation is illustrated in Figure 2.11. In order to proceed we need the relationship between the total percentage defectives, and the percentages of defectives at each end.

For our present purpose, put $L = 0$, $U = 1$, so that σ is represented in this scale, and we are interested in $0{\cdot}25 > \sigma > 0{\cdot}16$. Let a batch have a mean at x, then the proportion defective is

$$p = \Phi\left(-\frac{x}{\sigma}\right) + \Phi\left(\frac{x-1}{\sigma}\right) \quad (2.39)$$

and the two terms on the right of (2.39) are the proportions defective at the two ends. Equation (2.39) is shown in Figure 2.12; clearly the curve must be symmetrical about $x = (U + L)/2$. From a graph such as Figure 2.12, we can find the proportion defective at either end for a prescribed value of p.

Suppose now we require a plan which accepts a proportion $(1 - \alpha)$ of batches containing $100\theta_1$ % defectives, and which accepts a proportion β of batches containing $100\theta_2$ % defectives. From a graph such as Figure 2.12 we can find how proportions defective

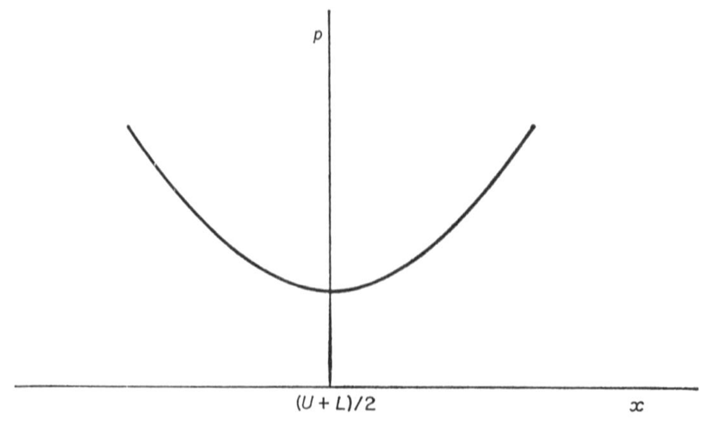

Figure 2.12. *Double specification limit: proportion defective.*

θ_1 and θ_2 are divided between the upper and lower limits by reading off x for $p = \theta_1$ and using the two terms on the right-hand side of (2.39) separately. Usually, the larger proportion defective θ_2 is affected very little by the presence of a double limit. Let the proportions be

$$\theta_1 = \theta_1' + \theta_1''$$
$$\theta_2 = \theta_2' + \theta_2''$$

and let $\theta_1' > \theta_1''$, $\theta_2' > \theta_2''$, so that usually θ_2'' will be negligible.

A sampling plan can now be constructed by combining two single specification plans using the parameters (α, θ_1'), (β, θ_2'), at each end.

The OC-curve can also be obtained from the single specification formula by transforming the percentage defective using (2.39).

(d) Single specification limit, σ unknown

When σ is unknown, our sampling plans must be modified by using either the sample standard deviation or the range to estimate σ. The theory becomes much more complex, and a number of papers have been written on it; see Owen (1967), Duncan (1974), and references therein. Here we give only a brief introduction.

If we have a single lower specification L to satisfy, we could use the criterion to accept if

$$(\bar{x} - L)/s > k, \tag{2.40}$$

and we have to choose k, and sample size n. Suppose we have OC-curve requirements $(\alpha_1\theta_1)$, $(\beta_2\theta_2)$, as before, then the probability statements we wish to satisfy involve the distribution of the statistic in (2.40), which has a non-central t-distribution. Tables of the non-central t-distribution are available, from which a sampling plan can be obtained (see Resnikoff *et al.* (1957)). A simple approximate procedure derives from writing (2.40) as

$$\bar{x} - ks > L$$

and we treat $(\bar{x} - ks)$ as approximately normal with mean $(\mu - k\sigma)$ and variance

$$\sigma^2\left(\frac{1}{n} + \frac{k^2}{2n}\right).$$

This approximate method leads to the formulae

$$\left.\begin{aligned} k &= (Z_\alpha Z_{\theta_2} + Z_\beta Z_{\theta_1})/(Z_\alpha + Z_\beta) \\ n &= \left(1 + \frac{k^2}{2}\right)\left(\frac{Z_\alpha + Z_\beta}{Z_{\theta_1} - Z_{\theta_2}}\right)^2 \end{aligned}\right\} \tag{2.41}$$

which are very similar to the σ known formulae (see Exercise 2.11.3).

(e) Double specification limit, σ unknown

Unfortunately this case is more complicated still, and we do not get an easy solution as we did in the σ known case.

Since σ in (2.39) is now unknown, there is a whole range of possible divisions of p to upper and lower limit defectives. From this we may deduce that the obvious procedure, based on

$$(\bar{x} - L)/s \geq k, \quad \text{and} \quad (U - \bar{x})/s \geq k,$$

does not have a unique OC-curve. In fact we have a band of OC-curves, but calculations have shown this band to be quite narrow.

Secondly, we may estimate the minimum percentage defectives as about

$$2\Phi\{(L - U)/2s\}$$

so that the procedure must reject if s is large. (Quite how large, depends on $\bar{x}$.)

However, the main calculations of a sampling plan can be related to the single specification plans. A satisfactory sampling plan is described in Duncan (1974), with illustrative calculations on the derivation.

(f) Defence sampling schemes for variables

A set of sampling plans for inspection by variables has been available for some time in the form of MIL-STD-414. This sampling scheme is based on precisely the same principles as for the attributes case, outlined in section 2.5. An AQL must be chosen, and assurance that this quality is met depends largely on the use of switching rules to and from tightened inspection.

The tables include (*i*) known standard deviation plans, (*ii*) unknown standard deviation plans, and (*iii*) average range plans, all for both single and double specification limits. These alternative sets of plans are matched to go through the same pair of points on the OC-curve. The two points specified are the AQL, at which the OC-curve varies between 0·89 and 0·99, and the proportion defective θ_2 at which the OC-curve is 0·10. The plans are indexed by AQL and sample size code letter so that it is easy to pick out three equivalent plans. The theory of Lieberman and Resnikoff (1955) is used in matching the plans.

MIL-STD-414 has not proved to be as popular as the parallel attributes schemes, and a review is given by Kao (1971). A draft British defence sampling scheme is now available, as BS6002 or DEF-STAN 05-30, and this scheme is largely a revision of MIL-STD-414. The fixed relationships between batch size and sample size have been improved, and some graphs have been used in the presentation, which are a great improvement. The underlying principles are essentially the same as those applying in MIL-STD-414.

(g) Decision theory approach

A decision theory approach to inspection by variables has been discussed by Wetherill (1959), Wetherill and Campling (1966), and Fertig and Mann (1974).

Suppose we take a measurement x on each item, and that the distribution of x in any batch is approximately normal with a mean μ and variance σ^2. If items are defective when $x < L$ say, we can specify utilities such as the following:

Decision	*Measurement*	*Utility*
Accept	$x > L$	k_1
Accept	$x < L$	$-k_2$
Reject	$x > L$	0
Reject	$x < L$	0

The percentage defective can be calculated for any given μ and σ^2. The theory now follows the attributes case, discussed in section 2.6. It is assumed that μ and possibly σ^2 vary from batch to batch in accordance with a prior distribution, which has to be specified.

The most important result from this approach is that the utilities of resulting sampling plans depend very strongly on a break-even percentage defective p_0, which turns out to be

$$p_0 = k_2/(k_1 + k_2).$$

No other parameter or assumption is as critical as p_0, and this again is parallel to the attribute case. The reader is referred to the references given above for details.

(*h*) *General discussion*

Although inspection by variables can lead to complicated theory, this cannot really be considered a disadvantage, since the theory can be reduced to sets of tables. However, there are more serious criticisms.

(*i*) It will be necessary to use a separate plan for each measured variable, whereas a single attributes plan could be operated.

(*ii*) The estimation of percentage defective is carried out assuming normality, and depends heavily on tail area properties. Thus the normality assumption is critical in estimating a small percentage defective.

(*iii*) A batch could be inspected and rejected without a single defective item having been spotted. A supplier might justifiably protest against this.

Against these criticisms, there are the advantages of better information on quality and usually great reductions in inspection effort.

It must be pointed out here, that if σ is known, an improved attributes plan can be worked. The definition of a defective is modified to, say

$$x < L + \Delta a,$$

leading to a much larger percentage of 'defectives'. This is called *increased severity testing*. For example, if $\sigma = 1$, $L = 0$ and $\mu = 2{\cdot}2$, the batch contains only $\Phi(-2{\cdot}2) = 1{\cdot}4\%$ of true defectives, whereas if we used increased severity testing with $\Delta = 0{\cdot}40$, the batch would be regarded as having $\Phi(-1{\cdot}8) = 4{\cdot}6\%$ of 'defectives'. This type of plan suffers from the disadvantage of other variables plans – that it depends rather heavily on the normality assumption for the relationship between true and 'increased severity' percentage defectives. However, a few calculations will show that it can lead to a much more efficient plan than an ordinary attributes sampling scheme; see Beja and Ladany (1974) for some work on this.

Exercises 2.11

1. Use the method of section 2.11(c) to obtain a sampling plan for Example 2.3, assuming $\sigma = 9$, and for $(\theta_1 = 0{\cdot}04,\ \alpha = 0{\cdot}05)$ $(\theta_2 = 0{\cdot}08,\ \beta = 0{\cdot}02)$.
2. Consider the case set out in section 2.11(b) above, and obtain the inspection for fraction defective plan which has the same producer's and consumer's risk points. (You may use the normal approximation method of Exercise 2.1.2.) Hence obtain a formula for the ratio of the sample sizes for the two methods of inspection.
3. Check the derivation of equations (2.41) as follows. (The method is parallel to that used in section 2.11(b).) The acceptance criterion is

$$\bar{x} - ks > L$$

which can be written

$$\frac{\bar{x} - ks - (\mu - k\sigma)}{\sigma\sqrt{(1 + \frac{1}{2}k^2)}\sqrt{(1/n)}} > \frac{L - (\mu - k\sigma)}{\sigma\sqrt{(1 + \frac{1}{2}k^2)}\sqrt{(1/n)}}.$$

We can now treat the left-hand side as approximately standard normal. Hence set n and k so that the probability of acceptance is $(1 - \alpha)$ when

$$(\mu - L)/\sigma = Z_{\theta_1}$$

and β when

$$(\mu - L)/\sigma = Z_{\theta_2}.$$

Also follow the method of section 2.11(b) to obtain an approximate OC-curve, by treating $(\bar{x} - ks)$ as approximately normal.

3. Control charts

3.1 Statistical quality control

In any production process, some variation in quality is unavoidable, and the theory behind the control chart originated by Dr W. A. Shewhart is that this variation can be divided into two categories, *random variation*, and variation due to *assignable causes*. Certain variations in quality are due to causes over which we have some degree of control, such as a different quality of raw material, or new and unskilled workers; we call these *assignable causes of variation*. The random variation is the variation in quality which is the result of many complex causes, the result of each cause being slight; by and large nothing can be done about this source of variation except to modify the process.

If data from a process are such that they might have come from a single distribution (frequently normal), having certain desired properties such as a mean in a specified range, the process is said to be *in control*. If on the other hand variation due to one or more assignable causes is present, the process is said to be *out of control*.

The Shewhart control chart is a simple device which enables us to define this state of statistical control more precisely, and which also enables us to judge when it has been attained. A sample of a given size is taken at frequent intervals, and a chart is kept of, say, the sample mean against time. The pattern of the points on the chart enables us to judge whether the process is or is not in control. There are different types of control chart depending on whether our measurement is a continuous variable, or whether we are measuring fraction defective, defects per unit, etc., but they are all rather similar, and we shall only go into details of the Shewhart control charts for variables; see section 3.2. In section 3.6 we briefly describe some other control charts.

As we shall see, the whole philosophy of a process either being in control or out of control is too naïve, but in many situations it is a sufficiently realistic model for our purposes. Little reliance can be placed on precise probability calculations, but instead we rely on a long history of widespread industrial applications in which it is found not only that the control chart works, but that it is of very great value. Some of the advantages will be discussed in the next section.

It is assumed, of course, that when there is evidence that assignable causes of variation are present, these causes can be traced and eliminated; this is usually the main aim of operating a quality control chart. Sometimes a quality control chart is used to search through past data, to see when changes of quality occurred.

3.2 Simple control charts for variables

The basic idea of a Shewhart control chart is very simple. Groups of, say, four items are drawn from a process at regular intervals, and measurements of quality made. Let us suppose that one variable is measured, and that we can assume this to have a normal distribution under a state of statistical control. The presence of assignable causes of variation would then show up in the data as variation outside the usual range for a normal variable. This can usually be attributed to a change in either μ or σ of the normal distribution for the measured variable. If, therefore, we calculate estimates of μ and σ from each of our groups of four, and plot these against time, the pattern of the points will enable us to detect whether the process is or is not in control.

Table 3.1 shows the results of hardness determinations on 25 samples of four titanium buttons, the samples being drawn at regular intervals from a chemical process (data by kind permission of I.C.I. Ltd, Mond Division). The graph of the group means shown in Figure 3.1(a) is called a control chart for means. In order to keep track of the process variability, we could plot the sample variances

$$s^2 = \sum_{1}^{4} (x_i - \bar{x})^2/3,$$

where x_i, $i = 1, 2, 3, 4$ are the observations for a group and $\bar{x}$ is the group mean. This is usually considered unduly complicated, and instead a graph is made of the group ranges, where

$$R = \text{range} = (\text{largest observation}) - (\text{smallest observation}).$$

A control chart for ranges is shown in Figure 3.1(b).

The range is an inefficient measure of dispersion if the group size is much more than 10. If the sample size is large, the efficiency can be

Table 3.1 *Hardness measurements of titanium buttons*

Set no.	*Hardness* (DPN)				*Mean* ($\bar{x}$)	*Range* (R)
1	125·8	128·4	129·0	121·0	126·1	8·0
2	125·2	127·0	130·4	124·6	126·8	5·8
3	121·8	126·8	127·0	129·8	126·4	8·0
4	131·0	130·0	127·2	127·0	128·8	4·0
5	128·6	122·8	125·4	126·4	125·8	5·8
6	122·0	123·8	131·2	121·8	124·7	9·4
7	122·8	129·8	126·2	128·8	126·8	6·4
8	120·2	130·0	125·6	144·0	130·0	23·8
9	124·8	123·8	130·2	128·8	126·9	6·4
10	127·0	126·4	122·2	129·0	126·2	6·8
11	131·8	127·6	123·8	123·2	126·6	8·6
12	129·8	125·6	128·2	127·6	127·8	4·2
13	127·6	125·6	128·2	126·8	127·1	2·6
14	124·2	122·8	124·8	124·6	124·1	2·0
15	125·4	129·4	123·6	127·2	126·4	5·8
16	130·8	122·8	125·4	126·2	126·3	8·0
17	127·4	131·0	123·0	122·8	126·1	8·2
18	124·8	122·6	122·8	123·6	123·5	2·2
19	123·8	130·0	128·4	130·0	128·1	6·2
20	128·8	141·2	138·8	136·2	136·3	12·4
21	126·4	123·8	128·8	129·6	127·2	5·8
22	130·8	127·4	126·0	125·2	127·4	5·6
23	129·6	128·4	123·2	125·8	126·8	6·4
24	124·4	127·0	130·0	122·8	126·1	7·2
25	129·2	126·2	128·0	123·2	126·7	6·0
				Totals	3175·0	175·6

improved by arbitrarily breaking the sample into groups of size 5–10, and then averaging the ranges of the groups. This may even be worth while doing for sample sizes between 10 and 20.

The action and warning limits drawn in on Figure 3.1 are an essential part of the control charts; the construction and use of the limits for the $\bar{x}$-chart is as follows. From past data of the process we obtain good estimates of μ and σ, and then we estimate the standard error of $\bar{x}$, which is $\sigma/\sqrt{n}$, where n is the sample size, that is four in our example. Then from tables of the normal distribution we find that if the process is in control, only one point in a thousand would be above $\mu + 3{\cdot}09\sigma/\sqrt{n}$, and only one point in a thousand below

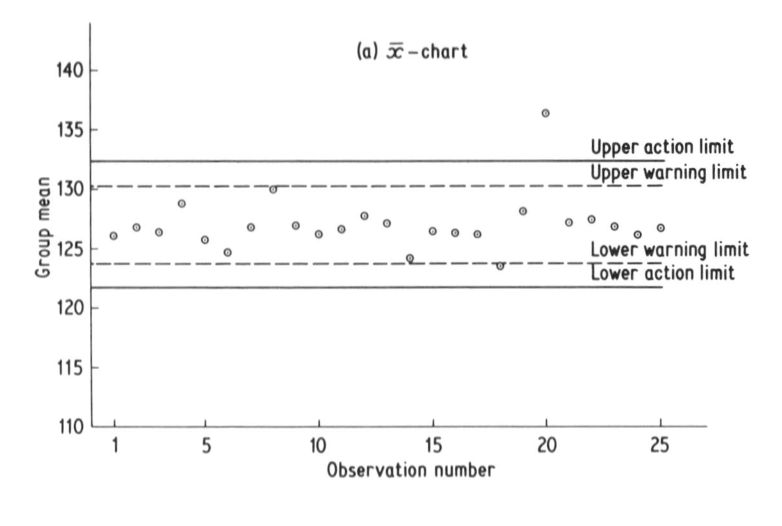

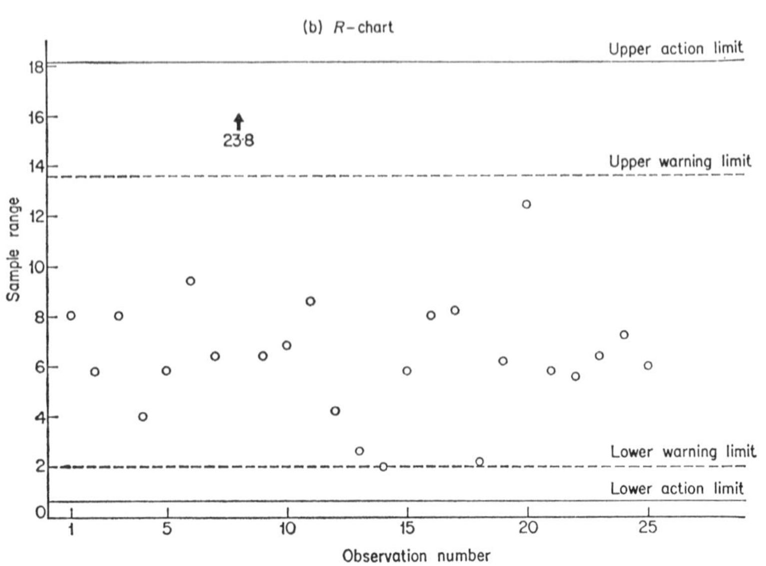

Figure 3.1. *Control charts for hardness of titanium buttons.* (*a*) *$\bar{x}$-chart,* (*b*) *R-chart.*

$\mu - 3{\cdot}09\sigma/\sqrt{n}$; these are the upper and lower action limits for the $\bar{x}$-chart. Therefore if a point falls outside the action limits, there is very strong evidence that assignable causes of variation are present, and action to trace and eliminate these causes should be initiated.

In order to improve the sensitivity of the control charts, warning limits are drawn in, as shown in Figure 3.1. These limits are usually set at $\mu \pm 1{\cdot}96\sigma/\sqrt{n}$, so that when the process is in control one point in forty should be above the upper warning limit, and one point in forty below the lower warning limit. Two successive points outside the warning limits are usually taken to be good evidence that assignable causes of variation are present.

Some people prefer to put the action and warning limits at $3\sigma/\sqrt{n}$ and $2\sigma/\sqrt{n}$ instead of the positions suggested above. There is very little difference between these two sets of positions, but the two traditions are well established.

The action and warning limits for the R-chart, Figure 3.1(b), are constructed on similar principles, but using the distribution of the range in normal samples. This results in the diagram being unsymmetrical. A short table of percentage points of the range is given in the Appendix tables.

Some people attempt to simplify the calculation of action and warning limits for an R-chart by calculating a standard error for range, and putting the action and warning limits at ± 3 and ± 2 standard errors. However, the simplification achieved is very slight, and anomalies arise such as negative lower limits. It seems better to use limits based on percentage points of the distribution of the range.

When setting up control charts, good estimates of μ and σ are required. At least 25 groups of observations should be obtained, and the mean and range of each group calculated. The means and ranges are then averaged, and the average of the means taken as an estimate of μ. The average of the ranges is converted to an estimate of σ by using Appendix Table 4. The position of action and warning limits can then be calculated. However, if the action and warning limits are drawn in, it may happen that some of the 25 groups used are out of control; such groups should be deleted and calculation of estimates of μ and σ repeated. If too many of the groups fall beyond the limits, this is evidence that the process is not in control, and limits cannot be calculated.

For the data shown in Table 3.1 this procedure works out as follows:

$$\text{Average range} = \bar{R} = 175{\cdot}6/25 = 7{\cdot}024;$$
$$\text{Average hardness} = 3175/25 = 127{\cdot}0;$$
$$\text{Estimate of } \sigma = 0{\cdot}4857 \times 7{\cdot}024 = 3{\cdot}41.$$

(The range factor for the estimate of σ is read from Appendix II table 4.) The one in a thousand action limits are at

$$\bar{x} \pm 3{\cdot}09\frac{\sigma}{\sqrt{n}} = 127{\cdot}0 \pm 3{\cdot}09 \times \frac{3{\cdot}41}{\sqrt{4}} = (121{\cdot}73,\ 132{\cdot}27),$$

and the warning limits are placed at

$$\bar{x} \pm 1{\cdot}96\frac{\hat{\sigma}}{\sqrt{n}} = 127{\cdot}0 \pm 1{\cdot}96 \times \frac{3{\cdot}41}{\sqrt{4}} = (123{\cdot}66,\ 130{\cdot}34).$$

Alternatively the positions of the action and warning limits for the $\bar{x}$-chart may be calculated directly from the average range by using the factors of Appendix II table 7. For groups of size four the factors are 0·7505 and 0·4760, so that we have

$$\text{Action limits:}\quad 127 \pm 0{\cdot}7507 \times 7{\cdot}024 = (121{\cdot}73,\ 132{\cdot}27)$$
$$\text{Warning limits:}\ 127 \pm 0{\cdot}4760 \times 7{\cdot}024 = (123{\cdot}66,\ 130{\cdot}34)$$

These numerical examples also illustrate the way in which Appendix II table 7 has been constructed, since for example, for samples of size four we have for action limits

$$3{\cdot}09 \times (\text{range factor for sample size } n)/\sqrt{n} = \frac{3{\cdot}09 \times 0{\cdot}4857}{2}$$
$$= 0{\cdot}7505.$$

If the action and warning limits are chosen to be at $\pm 3\sigma/\sqrt{n}$ and $\pm 2\sigma/\sqrt{n}$, the factors of Appendix II table 8 are used in place of the figures from Appendix II table 7. A similar and more extensive table of factors is given by Duncan (1974, Appendix table M).

For the range chart we read from Appendix II table 5 for the 0·1%, 2·5%, 97·5%, and 99·9% limits, sample size four, to obtain 0·20, 0·59, 3·98, and 5·31, and by multiplying by $\hat{\sigma}$ we have 0·68, 2·01, 13·57, and 18·10, which are the positions of the action and warning limits shown in Figure 3.1(b).

From Figure 3.1 we see that sets numbered 8 and 20 show evidence of lack of control. If the data of Table 3.1 were being used to set up quality control charts, these two sets should be omitted, and the boundaries recalculated. This is left as an exercise.

It is often found helpful to mark on the chart important changes in the process, such as a change of operator, a new batch of material, etc., as this may help to trace assignable causes of variation. It may also be necessary to recompute the action and warning limits from time to time, as either μ or σ changes.

There are several ways in which lack of control may be indicated by the control charts. There may be either a gradual change or a sudden change to a new value in either μ or σ, or both. Alternatively occasional points may fall out of control. Which of these possibilities occurs will help to indicate the underlying cause.

There are several rather arbitrary decisions in the above description. Firstly, we decided to track changes in μ and σ by plotting means and ranges; other statistics could be used, but these are the main ones. If the group size is much over ten the range is inefficient as a measure of variability, and the sample variance or standard deviation should be used.

Another decision is to select a group size, and the interval between selecting groups. Large group sizes drawn frequently would give good protection, but would cost a lot in inspection. The decision is therefore basically an economic one, and we shall discuss this again later. For a number of reasons, a group size of four or five is most common. The decision depends to some extent on properties of the control charts, which is a topic discussed in the next section.

The position of the warning and action limits is fixed by tradition and experience, but there is no reason why different limits should not be used in a particular application.

Exercises 3.2

1. Recalculate the positions of the action and warning limits for Table 3.1 data after sets 8 and 20 have been omitted. Are there now any further sets outside the limits?

2. Calculate the positions of action and warning limits of $\bar{x}$- and R-charts for the same process as Table 3.1 data but suppose that it was decided to take measurements (*a*) 8 at a time, and (*b*) 12 at a time.

3. When setting up $\bar{x}$- and R-charts it was found that for groups of size five

$$\text{Overall mean} = 54.2, \quad \text{Mean range} = 6{\cdot}714.$$

Calculate the position of the limits.

4. It has been decided to install quality control charts on a relatively new batch process. Three batches are produced each shift, and each batch is assessed by a boiling point determination. Owing to a change in plant operating procedure, only the results for the last 75 batches are relevant to present behaviour and the boiling points of these batches are listed in Table 1. Construct quality control charts for boiling point using these data, and plot the subsequent results given in Table 2. Assuming that when disturbances were found they were promptly investigated and dealt with, on which occasions should investigations have been made? Is there any indication that the control charts should be amended?

Table 1 *Original boiling point data*

Shift No.	*Boiling point* (°C)		
1	46·0	45·5	46·5
2	44·6	47·3	46·0
3	45·9	44·5	46·2
4	44·7	43·9	45·5
5	45·0	45·5	45·6
6	43·9	46·2	43·6
7	45·3	45·9	45·5
8	44·8	45·4	46·7
9	44·2	44·5	43·8
10	44·6	45·0	46·0
11	45·4	44·7	45·3
12	45·2	45·3	46·1
13	44·8	46·2	44·4
14	48·3	47·9	48·1
15	45·5	47·0	45·4
16	44·9	47·0	46·5
17	47·0	46·2	44·7
18	44·4	47·0	39·9
19	42·3	46·5	44·1
20	44·1	46·3	47·5
21	45·4	46·6	44·1
22	45·6	45·5	47·5
23	47·0	47·8	44·1
24	45·1	46·4	48·1
25	46·3	45·5	45·0

Table 2 *Subsequent boiling point data*

Shift No.	*Boiling point* (°C)			*Mean* ($\bar{x}$)	*Range* (R)
26	43·7	46·6	45·7	45·3	2·9
27	44·8	44·9	46·1	45·3	1·3
28	46·7	44·8	45·6	45·7	1·9
29	44·8	45·9	46·7	45·8	1·9
30	45·5	44·1	45·1	44·9	1·4
31	46·4	45·7	44·8	45·6	1·6
32	45·2	45·9	45·2	45·4	0·7
33	45·3	46·9	46·0	46·1	1·6
34	44·7	45·4	45·8	45·3	1·1
35	47·1	45·3	44·6	45·7	2·5
36	42·5	45·8	43·7	44·0	3·3
37	43·2	44·0	44·2	43·8	1·0
38	44·3	45·6	43·5	44·5	2·1
39	45·6	44·2	44·7	44·8	1·4
40	47·1	45·9	44·1	45·7	3·0
41	46·3	43·7	46·2	45·4	2·6
42	42·0	47·1	48·1	45·7	6·1
43	47·7	44·1	47·1	46·3	3·6
44	44·9	46·5	44·1	45·2	2·4
45	45·1	44·3	45·1	44·8	0·8
46	44·6	45·3	45·8	45·2	1·2
47	45·5	45·2	46·4	45·7	1·2
48	47·0	44·5	46·5	46·0	2·5
49	45·4	45·7	49·9	47·0	4·5
50	46·1	47·6	46·7	46·8	1·5
51	47·3	46·3	46·5	46·7	1·0
52	48·1	47·8	46·0	47·3	2·1
53	48·3	47·6	47·5	47·8	0·8

3.3 Properties of the charts

(a) The $\bar{x}$-chart

It will be natural to consider the properties of $\bar{x}$-charts and R-charts in terms of the OC-curve (see section 2.1) or the ARL function (see section 2.2). If only a single pair of limits is used at $\pm 3{\cdot}09\sigma/\sqrt{n}$ on an $\bar{x}$-chart, and we consider only changes in the mean, then the OC-curve is

$$P(\theta) = \Phi(3{\cdot}09 - \theta\sqrt{n}/\sigma) - \Phi(-3{\cdot}09 - \theta\sqrt{n}/\sigma), \qquad (3.1)$$

where θ is the deviation of the mean of the process from the target.

Usually, one of the terms in this expression is small or negligible. The ARL function is $1/(1 - P)$, in terms of numbers of groups, and is more directly meaningful in that it tells us how much product, on average, passes before a change in the mean is detected.

The ARL function of the $\bar{x}$-chart tends to be rather flat. In order to sharpen it up, extra rules for taking action have been considered, such as the following.

(*i*) As indicated in section 3.2, warning limits are drawn at $\pm 1{\cdot}96\sigma/\sqrt{n}$, and action taken if two consecutive points fall outside these.

(*ii*) Action is taken if K out of the last N points fall between the action and warning limits.

(*iii*) A third set of lines is drawn at $\pm\sigma/\sqrt{n}$, and action taken if three consecutive points fall outside these.

There are other more complicated rules: see Page (1961), Grant and Leavenworth (1972, pp. 97–8), and Roberts (1966), for a brief discussion of these. Rule (*i*) above is used quite frequently, but the others tend to remove one of the chief advantages of Shewhart charts – simplicity. Furthermore, these rules do not sharpen up the ARL function very much; this can only be done by bringing the action and warning limits in towards the mean.

So far, we have assumed that σ in (3.1) is constant. However, both changes in the process mean and variance are regarded as indications that the process is out of control, and (3.1) should be regarded as a function of θ and σ. Figure 3.3 shows a contour map of the ARL function of the $\bar{x}$-chart, when just the outer action limits are used. The OC-function can be plotted in a similar way, and is left as an exercise, see Exercise 3.3.1. It is obvious from Figure 3.3 that the $\bar{x}$-chart alone gives very little protection against variations in σ, which is the reason why an R-chart or s^2-chart must accompany the $\bar{x}$-chart.

It is important to bear in mind here that the point of calculating the OC-curve or ARL function is to assist in choosing a group size n, and the interval t between taking groups. The actual ARL for a time interval t is $t/\{1 - P(\theta, \sigma, n)\}$. Now for a given value of θ, we may be able to fix an upper limit to the ARL we would desire for our plan, and let this value be m_θ, so that

$$m_\theta = t/\{1 - P(\theta, \sigma, n)\},$$

or

$$P(\theta, \sigma, n) = (1 - t/m_\theta). \tag{3.2}$$

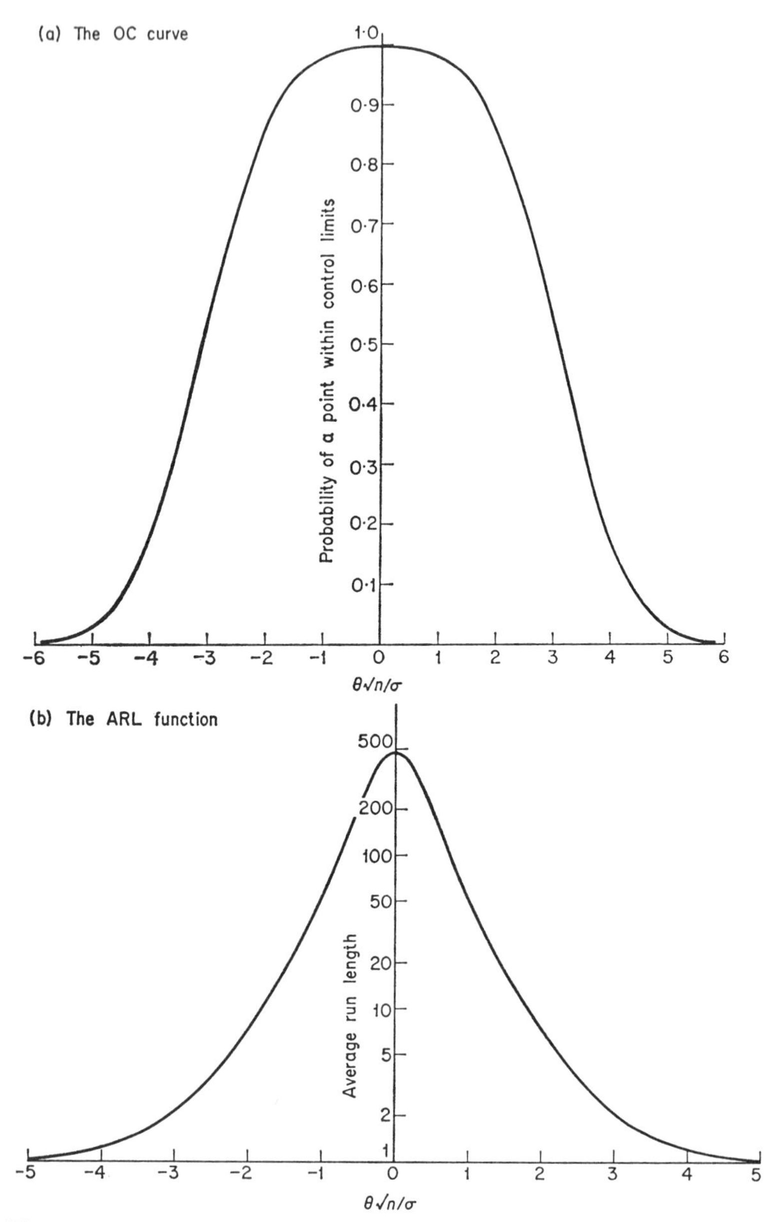

Figure 3.2. *The OC-curve and ARL function of the $\bar{x}$-chart. (a) The OC-curve. (b) The ARL function.*

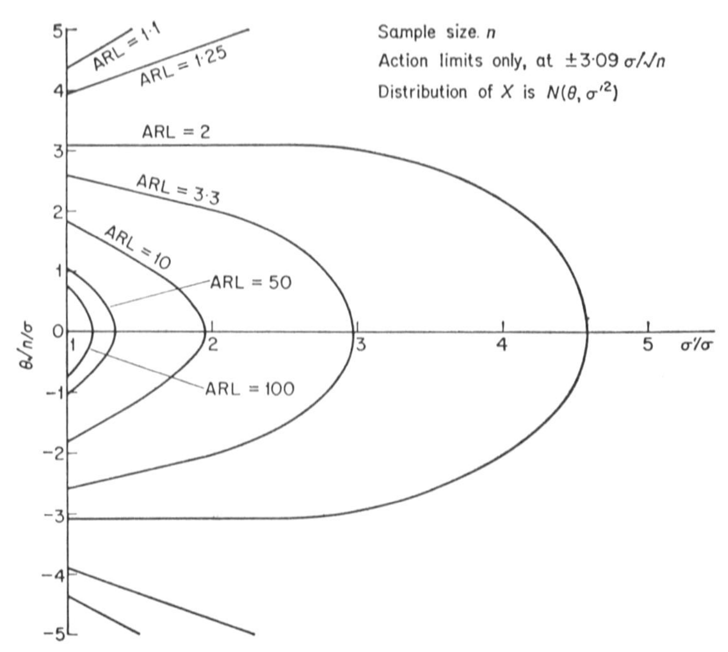

Figure 3.3. *The ARL function of the $\bar{x}$-chart.*

But $P(\theta, \sigma, n)$ is a function only of $\theta\sqrt{n}/\sigma$, and a graph of $(\theta\sqrt{n}/\sigma)$ against (t/m_θ) will be essentially the OC-curve. From such a graph a series of possible values of t and n will be clear. A final selection can be made on the basis of the ARL when $\theta = 0$.

In many applications of quality control charts, samples of size 4 or 5 are used. The reasons for this are partly that it is usually better to take a smaller sample size frequently than a larger sample size less frequently. A further point is that a small sample size keeps down the amount of arithmetic necessary for plotting.

Clearly, a quality control scheme is aiming at an economic balance between the costs of inspection and the costs of passing bad material. This economic structure can be formalized and studied in a way rather parallel to the decision-theory approach to sampling inspection plans outlined in Chapter 2. Such a theoretical study was undertaken by Duncan (1956), and this will be described in the next section.

It should be emphasized that the discussion given above of the OC-curve and ARL-function of the $\bar{x}$-chart has only assumed action limits. Page (1955) obtained the ARL-function when warning limits are used in addition to action limits.

In order to demonstrate the result, let us adopt the following notation:

p_0 = probability that a point lies between the warning limits;

p_1 = probability that a point falls between the warning and action limits;

L_0 = ARL given the first point lies between the warning limits;

L_1 = ARL given the first point lies between the warning and action limits.

We assume that all sample means are independently and normally distributed with expectation μ, and variance σ^2/n, where n is the group size. Thus p_0, p_1, L_0, and L_1 are all functions of μ and σ. If we now consider what happens if one more mean is observed, we obtain the equations

$$L_0 = 1 + p_0 L_0 + p_1 L_1$$
$$L_1 = 1 + p_0 L_0$$

where unity accounts for the extra mean observed. By solving these equations for L_0 we obtain

$$L_0 = (1 + p_1)/(1 - p_0 - p_1 p_0)$$

which is the ARL function with warning limits.

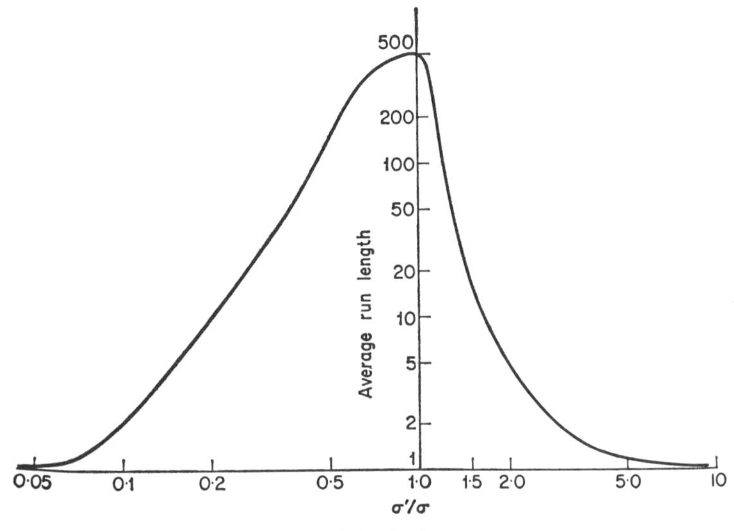

Figure 3.4. ***The ARL function of the R-chart.***

(b) The R-chart

The OC-curve or ARL function of the R-chart is independent of the process mean μ, and depends only on σ, so that they are rather easier to study than for the $\bar{x}$-chart. If σ' is the true variance of the process, the distribution of R/σ' is independent of σ', and given in tables. If therefore an R-chart is used with just upper and lower action limits R_U, R_L, it is merely necessary to look up the tables to find the probability that $(R/\sigma') > R_U$ or $(R/\sigma') < R_L$. In this way the graph shown in Figure 3.4 can be obtained.

Similar considerations now arise to those given above for the $\bar{x}$-chart. Extra rules for detecting lack of control can be used. The choice of a group size and inspection interval can be chosen on ARL considerations, etc.

(c) General

The choice of the group size and inspection interval is a very important aspect of quality control. In making the choice, considerations from both the $\bar{x}$-chart and the R-chart should be borne in mind, and a compromise made which will be reasonably satisfactory for both. However, in some situations, a process will be much more likely to go out of control through changes in one of the parameters μ and σ rather than both, so that one of the charts will be much more important.

This leads on to a rather unsatisfactory feature of the above theory. When a process goes out of control, it is assumed that the distribution of measurements on the items is still normal, but with a new μ or σ (or both). In fact it is quite likely that the distribution may become very non-normal, perhaps bimodal, or severely skew. The theory should therefore be used as a guide only. Experience shows that it is a sufficiently realistic model in many practical situations. Clearly, for any precisely defined model for deviations of the process from control, OC-curves and ARL functions could in principle be calculated.

(d) Some advantages of quality control

The Shewhart control charts are used in a manufacturing process for ‘trouble shooting’. The charts indicate when there is trouble, and from the form of the $\bar{x}$-chart and R-chart, and from rough indications of when the trouble started, it may be possible to obtain a good guide to the source of the trouble.

In practice it has been found that the charts are valuable in indicating when to leave a process alone as well as when to take action.

Sometimes the charts have revealed that operators attempting to correct a process have actually been an assignable cause of quality variation.

There are also some useful by-products of a quality control scheme. The continuous set of data on product quality which the scheme provides can be used in a number of ways. A periodic assessment of departmental performance could be made. A comparison of the quality delivered by different suppliers could also be made; it could also lead to an improvement of inspection standards, and greater quality conciousness on the part of the production team.

In many factories, Shewhart control charts are being superseded by CUSUM schemes, and these are discussed in the next chapter. Advantages such as those described above also apply to these schemes.

Exercises 3.3

1. Plot the OC-function of the $\bar{x}$-chart, when both μ and σ may vary.
2. Draw up a table of $\theta\sqrt{n}/\sigma$ against t/m_θ, from (3.2).
3. For an $\bar{x}$-chart with action and warming limits set at $3{\cdot}09\sigma/\sqrt{n}$ and $1{\cdot}96\sigma/\sqrt{n}$ respectively, calculate the ARL when $n = 4$, $\sigma = 2$, and the process mean differs from target by one unit.
4. Find suitable values for the sample size and sampling interval for an $\bar{x}$-chart so that the ARL (calculated in terms of items produced) is 50,000 when $\theta = 0$, and 150 when $\theta = 2$. Assume that $\sigma = 1$, and that 10 items are produced per minute.

3.4 The economic design of $\bar{x}$-charts

The method of designing an $\bar{x}$-chart suggested in the previous section is rather unsatisfactory since the design problem is basically an economic one. An interesting approach to the design of $\bar{x}$-charts which incorporates costs is given by Duncan (1956) (see also Chiu (1973)); the analysis leads to some important qualitative results which should be borne in mind, whether or not one wishes to use the method in a practical case.

A sample of size n is drawn from a process every h hours and an $\bar{x}$-chart plotted. If the process is in control, the distribution of $\bar{x}$ is $N(0, \sigma^2/n)$, and if the process is not in control the distribution of $\bar{x}$ is still normal with variance σ^2/n, but with a mean of either $+\delta\sigma$ or $-\delta\sigma$. Action limits are drawn on the $\bar{x}$-chart at $\pm k\sigma/\sqrt{n}$, and a search is made for an assignable cause if and only if a point falls beyond the action limits. The problem is to choose the parameters (h, k, n) to minimize the costs, as described below. The parameter δ is considered fixed at a value equal to the sort of deviation it is considered that the plan should detect.

The process is assumed to start in control, and the time until an assignable cause changes the mean to $\pm\delta\sigma$ has a distribution $\lambda e^{-\lambda t}$, so that the average time interval to an assignable cause is $1/\lambda$. The time taken to plot the points on the chart is taken to be gn, and the average time taken to check and eliminate an assignable cause is D.

If an assignable cause occurs within an interval, the average time at which it occurs is

$$\frac{\int_0^h t\lambda e^{-\lambda t}\, dt}{\int_0^h \lambda e^{-\lambda t}\, dt} = \frac{1 - (1 + \lambda h)e^{-\lambda h}}{\lambda(1 - e^{-\lambda h})}$$

$$\simeq \frac{h}{2} - \frac{\lambda h^2}{12} + 0(\lambda^3 h^4), \tag{3.3}$$

and this approximation will be valid if λ is small and h moderate.

The probability of detecting an assignable cause is

$$P(k, n, \delta) = \Phi(-k - \delta\sqrt{n}) + \Phi(-k + \delta\sqrt{n}). \tag{3.4}$$

The probability of a point outside the action limits when the process is in control is

$$P(k, n, 0) = \alpha, \quad \text{say.}$$

The probability that an assignable cause is detected only on the rth sample after it has occurred is $P(1 - P)^{r-1}$, so that the average number of samples taken is $1/P$. We can therefore define the average cycle length as

$$\begin{aligned} C = {} & \text{(average time in control)} \\ & + \text{(average time out of control until detection)} \\ & + \text{(plotting delay)} + \text{(checking delay)}, \end{aligned}$$

so that

$$C = \frac{1}{\lambda} + \left(\frac{h}{P} + \frac{\lambda h^2}{12} - \frac{h}{2}\right) + gn + D. \tag{3.5}$$

The proportion of time the process is in or out of control respectively can therefore be written

$$\beta = 1/C\lambda, \quad \gamma = 1 - \beta. \tag{3.6}$$

The frequency of assignable causes per hour is simply $1/C$. The expected number of false alarms per hour is

$$\frac{\text{the expected number of false alarms while process in control}}{\text{average cycle length}}$$

$$= \frac{\alpha}{C} \sum_{i=0}^{\infty} \int_{ih}^{(i+1)h} i\lambda e^{-\lambda t}\, dt = \alpha e^{-\lambda h}/\{(1 - e^{-\lambda h})C\}$$

$$\simeq \beta\alpha/h \tag{3.7}$$

We are now ready to define costs:

T = The cost of searching for an assignable cause when none exists.

W = the cost of checking and eliminating an assignable cause.

b = the cost per sample of plotting on the $\bar{x}$-chart.

c = the cost of measuring, per item.

M = the loss in profit when the process is out of control.

The total average loss L per unit time therefore arises from four terms,

L = (loss from process when out of control)
+ (cost of operating the $\bar{x}$-chart)
+ (cost of an assignable cause)
+ (cost of false positives).

This is
$$L = M\gamma + b/h + cn/h + W/C + T(\beta\alpha/h). \tag{3.8}$$

This can be written

$$L = \frac{\lambda W + \lambda MB + \alpha T/h}{1 + \lambda B} + \frac{(b + cn)}{h} \tag{3.9}$$

where
$$B = \lambda C - 1 = \left(\frac{1}{P} - \frac{1}{2} + \frac{\lambda h}{12}\right)h + gn + D.$$

A minimum of L exists for choice of (h, k, n) but further approximations are necessary in order to obtain simple answers. By assuming α and λ both small, Duncan obtains the following equations:

$$h \simeq \sqrt{\left\{\frac{\alpha T + b + cn}{\lambda M(1/P - \frac{1}{2})}\right\}} \tag{3.10}$$

$$(\alpha T + b)/c = P^2(1/P - \tfrac{1}{2})\frac{\partial n}{\partial P} - n, \tag{3.11}$$

$$e^{-k^2/2} = \sqrt{(2\pi n)}c/\delta T. \tag{3.12}$$

A numerical examination of the solutions yields the following conclusions.

(*i*) The optimum sample size is largely determined by δ, the size of the shift which we wish to detect. For $\delta = 2$ or more, an n in the range 2 to 6 will usually be optimum, but for smaller δ much larger sample sizes should be used.
(*ii*) The choice of the interval between samples, h, is affected mostly by the value of M; the larger M, the smaller h.
(*iii*) The value of k is determined largely by T and W; large values of T and W lead to large values of k.

In particular therefore the analysis is important in revealing the situations in which the standard choices of k and n are not appropriate. Further results would be of interest, such as an investigation into how sharp or flat the optimum is to the parameters of the model.

Duncan (1971) discusses an extension of the above work to the case of several assignable causes, and hence several δ_j. He demonstrates that a near optimum solution can be obtained by using the single assignable cause model. An algorithm for computer determination of the design parameters of an $\bar{x}$-chart is given by Goel, Jain, and Wu (1968).

A simplified semi-economic approach to the design of $\bar{x}$-charts following the general outline of Duncan's work is given by Chiu and Wetherill (1974). One point on the OC-curve can be set arbitrarily, and this is the probability that a point lies beyond the control limits, given that the mean has deviated by a specified amount from the target value. Tables are provided by which the parameters n, h, and k can be determined, which minimize the costs subject to this probability restriction. It is demonstrated that in general the resulting plan is close to the exact minimum cost plan.

Exercises 3.4

1*. Refer to Duncan's paper and check the working from (3.9) to (3.10), (3.11) and (3.12).
2*. Examine the difficulties in carrying through the above analysis if a probability distribution is assumed for δ. (Some alteration will have to be made to the cost of checking and eliminating an assignable cause, W, to make this depend on δ.)
3*. Consider the effect of introducing warning limits on the analysis of this section.

3.5 Specifications for variables

So far in our discussion of control charts, we have not mentioned any externally specified limits (or tolerances) within which the measurements must lie. The reason for this is that the Shewhart control chart is based on the concept of statistical control, and, for example, a process mean is controlled by using limits derived using the *observed variation* of the process. Externally applied limits are considered as largely irrelevant. In discussing the effect of specifications we shall refer to two case histories discussed in detail by Grant and Leavenworth (1972).

Example 3.1 (Grant and Leavenworth, 1972, pp. 16–22). The limits on the pitch diameter of threads on some aircraft fittings were specified as $0{\cdot}4037 \pm 0{\cdot}0013$ inch. Analysis of five measurements per hour for twenty hours established that the process variability was such that the specified limits could be met reasonably easily. (The process standard deviation is about 0·0003 inch.) □ □ □

Example 3.2 (Grant and Leavenworth, 1972, pp. 22–7). A rheostat knob was produced by plastic moulding using a metal insert purchased from another manufacturer. A critical dimension of the knob was given specification limits of $0{\cdot}140 \pm 0{\cdot}003$ inch by the engineering department. The analysis of some data showed that the process was evidently in control, but that the standard deviation of the process itself was slightly more than 0·003 inch. Therefore, even if the process was in control with a mean exactly at 0·140 inch, a large proportion of the production would be considered defective. Very little could be done to reduce the process variability, which was partly due to the metal insert, but an examination of the tolerances showed them to be much more narrow than was necessary and that limits of 0·125 inch to 0·150 inch would be satisfactory. □ □ □

Let us assume that we have a process in control and that its standard deviation can be estimated; different control charts are then applicable according to the value of the ratio

$$\text{(total specification tolerance)/(standard deviation).} \qquad (3.13)$$

If this ratio is small, say less than 6, then more than 0·2% of defective items will be produced even if the process mean is held exactly on the target value; the smaller the ratio, the larger the minimum proportion of defective items produced. This was found to be the situation initially in Example 3.2. Two lines of approach are open:

(*i*) attempt to reduce the process variability, or

(*ii*) examine whether the tolerances have been set much more narrowly than is necessary. (Sometimes on checking what was meant by the tolerances, one finds that 5% limits were implied.)

If neither of these approaches is successful, the best we can do is run a control chart and sift out the defectives by inspection.

If the ratio (3.13) is moderate, say between 6 and 10, the ordinary control charts which were described earlier in this chapter should be used.

The final case is when the ratio (3.13) is large, so that it is no

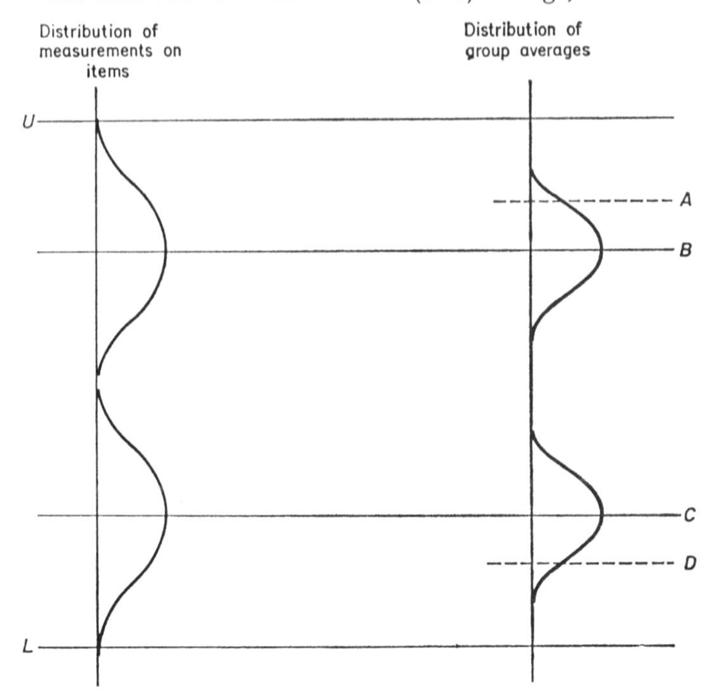

Figure 3.5. *Standard derivation of modified control limits.*

longer necessary to control the process within the narrow limits described earlier, and we can use 'modified control limits'. An important paper on the theory of modified control limits is Hill (1956), in which the author rejects the 'standard approach' and suggests an alternative. First we shall briefly describe the standard theory.

Let the specified tolerances be U and L as shown in Figure 3.5. If the mean of the process is at B, then only 0·1% of the items will have measurements above U, and we may consider the upper specification to be met. Following the method given in section 3.2, evidence that the process mean is above B is shown by a group average above A, where AB is $3{\cdot}09\sigma/\sqrt{n}$. (Some people use a distance $1{\cdot}96\sigma/\sqrt{n}$.) The lower limit follows similarly, and the modified control limits are A and D. If the specified tolerance UL is only $6{\cdot}18\sigma$ wide, the lines B and C are identical, and the modified limits reduce to the ordinary action limits of section 3.2.

The main fallacy in this argument is that the original action limits were based on the concept of statistical control, whereas with modified control limits, the process mean is allowed to wander, and we no longer have control. Thus in the Shewhart control chart, a point near (but inside) the action limits is taken as evidence that the process is still in control, and the true mean is less than the observed mean. With modified control charts we have no basis for assuming that the true mean is at all less than the observed mean. Therefore by using the modified limits of Figure 3.5, the process mean would be allowed to rise above B without any action being deemed necessary, and a considerable proportion of defectives could be produced; see Exercise 3.5.1.

It is somewhat surprising that the modified limits are placed outside the limiting positions for the process mean. By doing so we arrive at the paradox that by increasing the group size, n, the modified limits would be placed further away from the tolerances.

A further objection to the theory outlined above is that it depends heavily on normality. The Shewhart $\bar{x}$-chart is a technique for controlling a mean, and no assumptions are made about the tolerances satisfied by individual items. The above theory in contrast depends rather heavily on tail area probabilities of the normal distribution, and so is sensitive to the normality assumption.

Hill (1956) points out that many authors have recognized the objections to the standard approach to modified limits, and stressed the need for extra caution. Hill suggested that the modified limits should be placed so that if the process mean reached the positions B or C in Figure 3.5, there was only a 5% probability of *not* taking action. This leads to placing the modified limits at a position $1{\cdot}645\sigma/\sqrt{n}$ *inside* B and C. The width of these modified limits is therefore

$$(U - L) - 2(3{\cdot}09 + 1{\cdot}645/\sqrt{n})\sigma.$$

If these limits are narrower than ordinary limits, that is, less than $6{\cdot}18\sigma/\sqrt{n}$, then the best we can do is to use the ordinary limits. Therefore we use ordinary limits whenever

$$(U - L)/\sigma < (6{\cdot}18 + 9{\cdot}47/\sqrt{n}). \qquad (3.14)$$

One difficulty with the ordinary Shewhart control chart is that it is implied that it is practicable to trace and eliminate all causes of variations in the process mean (or variability) except the purely random variation. The idea is rather that a foreman can just 'twiddle a knob' and put things right. Today manufacturing processes are becoming increasingly complicated and more extensively automated, and this simple approach is frequently untenable. For example, in

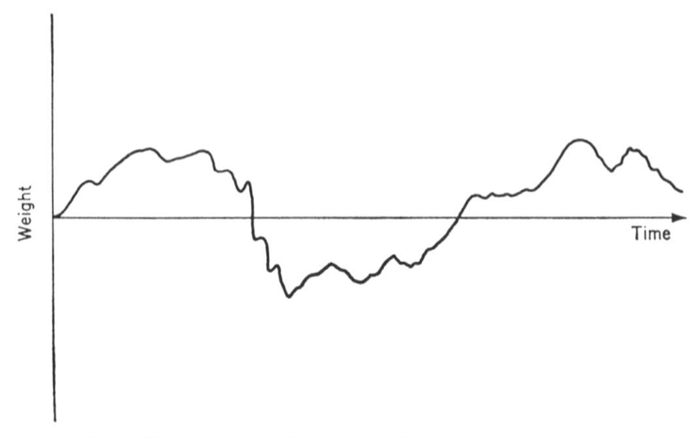

Figure 3.6. *Mean weight of chocolate bars.*

the production of chocolate bars, it is not possible to control the mean weight in the Shewhart sense. The mean weight seems to wander, rather like a first-order autoregressive process (see Appendix I), but with assignable causes of variation superimposed; see Figure 3.6. (These variations are very small.) This sort of behaviour has been observed by the author in a number of cases, and frequently either little is known about the causes of the oscillations of the process mean, or else very little can be done about them. By comparison with the within-group variation, even short-term variations of the mean may prove to be statistically significant, but often of little practical importance. In these situations a 'modified control limit' approach is desirable with sufficient warning being given of the process mean approaching the tolerances for action to be taken. The method suggested by Hill should therefore be of increasing practical importance today.

A further complication which we have avoided so far is when there is appreciable error of measurement in the observations. Some of the points arising are brought out in a paper by Desmond (1954) dealing with inspection of voltage regulators for private motor-cars.

The voltage regulators are required to operate in the range 15·8 to 16·4 volts. At the end of the production line the regulators are inspected, and any not working within the tolerance range are passed back for resetting. Even under good conditions a large proportion of output had to be reset, and an experiment was designed to investigate the sources of error. This experiment revealed a very large measurement error, the component of variance due to this being 0·0511, equal to a standard deviation of 0·226. This error of measurement was superimposed on the natural variation of true regulator operating voltages which had a variance of about 0·0435 (standard deviation 0·209). This made a total variance of $0{\cdot}0511 + 0{\cdot}0435 = 0{\cdot}0346$, equal to a standard deviation of 0·308.

Therefore, if an ordinary Shewhart control chart is operated centred at 16·1 volts, the action limits are placed at

$$16{\cdot}1 \pm 3{\cdot}09 \times 0{\cdot}308 = 15{\cdot}15,\ 17{\cdot}05 \text{ volts.}$$

However, provided the chart shows that production is in control, it can be assumed that 99·8% limits for the true regulator voltage measurements are

$$16{\cdot}1 \pm 3{\cdot}09 \times 0{\cdot}209 = 15{\cdot}45,\ 16{\cdot}75.$$

(If operating voltages outside this range occur, then a measurement error would sooner or later give a point beyond the outer control limits.) A Shewhart control chart was therefore run, with extra warning lines, and eventually the variances reduced, so that most of the production was within the desired limits.

The above argument sounds rather hazardous, and extra precautions were taken. Desmond's paper is very clear, including photographs of the physical set-up, and is well worth reading.

The problem of components of variance in a quality control situation is often overlooked, and demands a deeper investigation than anyone seems to have given it.

Exercises 3.5

1. Consider the standard approach to modified limits given in Figure 3.5, with $\sigma = 1$, $n = 4$, and $U - L = 10$. Calculate the ARL

when the process mean is at $(B + \theta)$, for $\theta = 0$, 0·5, 1·0.

2. Construct a short table of $(3{\cdot}09+1{\cdot}645/\sqrt{n})$ and $(6{\cdot}18+9{\cdot}47/\sqrt{n})$, for use with Hill modified limits, for $n = 2$ (1) 10.

3. Consider how to insert warning limits on to the Hill modified limits.

4*. Formulate and examine a model for the economic design of modified control limits.

3.6 Control charts for qualitative data

In this section we give a very brief account of two other types of control chart in current use, and we refer the reader to more extensive accounts elsewhere. The principles of derivation and operation of the charts are very similar to those for control charts for variables, discussed above.

(a) Control chart for fraction defective

In a complicated assembly, many variables may be measured, and in principle control charts could be kept for each one. An alternative is to note simply whether the item is effective or defective as in Example 3.3.

Example 3.3 (Armstrong, 1946). A foundry is continuously producing side frames of railway-cars, and a random sample of 50 of each day's output is inspected, and the number rejected is noted. □□□

Samples may be drawn every hour, day, etc., or sometimes there is 100% inspection, and we simply observe the number of defectives. If the sample size is constant, then either the number of defectives or the fraction defective is plotted serially, in the usual way; see Figure 3.7.

The action and warning limits should be drawn in using calculations based on the binomial distribution. However, the sample size is usually large enough for the normal distribution to be used as an approximation. The standard errors of the number of defects and fraction defective are respectively

$$\sqrt{\{np(1-p)\}} \quad \text{and} \quad \sqrt{\{p(1-p)/n\}},$$

where p is the long-run average fraction defective. Action and warning limits are now drawn in at $\pm 3{\cdot}09$ and $\pm 1{\cdot}96$ standard errors, as before. It will be necessary to recalculate the limits periodically, especially when a chart is only recently started.

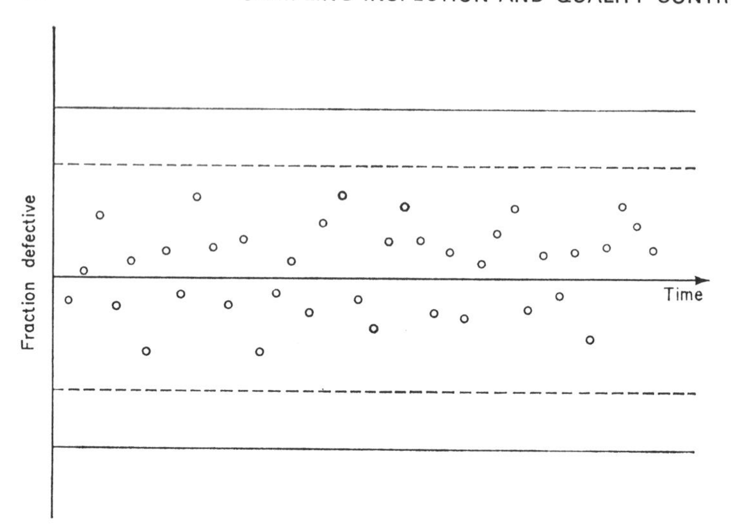

Figure 3.7. *Control chart for fraction defective.*

If the sample size varies, then either we must plot the fraction defective and recalculate the limits each time, or else plot

$$\frac{p_i' - p}{\sqrt{\{p(1 - p)/n_i\}}} \tag{3.15}$$

where p_i' is the fraction defective of the ith sample, and n_i is the sample size. The quantity (3.15) should be approximately standard normal, so that action and warning limits should be at $\pm 3{\cdot}09$ and $\pm 1{\cdot}96$ respectively.

For further information on fraction defective charts, and worked examples, see Duncan (1974), Grant and Leavenworth (1972), Huitson and Keen (1965), or British Standard 1313 (1947). An economic approach to the design of fraction defective charts is given by Chiu (1975).

(b) Control chart for number of defects

In the inspection of complex assemblies such as motor vehicles, aircraft wings, or even in observations such as the number of surface defects in a metal sheet, our data can be reduced simply to the number of defects per item. Data of this kind are often well fitted by the Poisson distribution, although this is not necessarily the case if the density of defects is heterogeneous. A control chart could be constructed using exact probability calculations from the Poisson distribution, but it is usually satisfactory to treat the number of defects, c, as an approximately normal variable with a mean m and variance $\sqrt{m}$, where m is the long-run average number of defects.

A control chart is operated by selecting a group of n items randomly at preselected time intervals, and counting the total number of defects in the group, $c_1, c_2, \ldots$. The observed values c_i are then plotted on a chart (Figure 3.8) and action and warning lines put in at

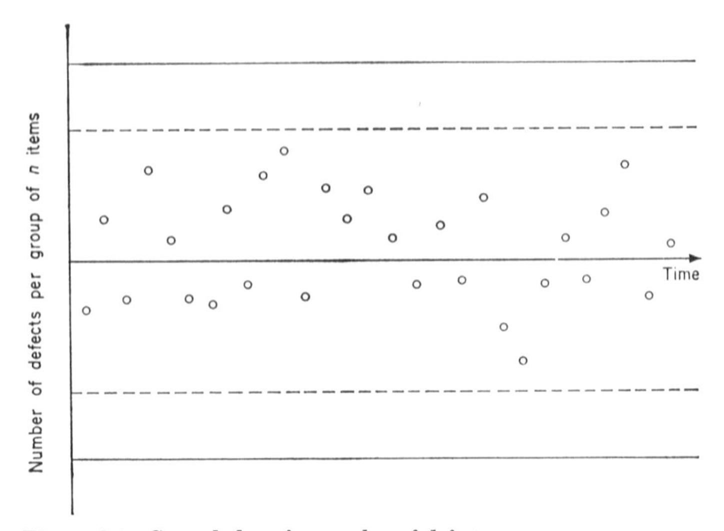

Figure 3.8. *Control chart for number of defects.*

$\pm 3{\cdot}09\sqrt{m}$ and $\pm 1{\cdot}96\sqrt{m}$ respectively, as in the construction of control charts for fraction defective, Figure 3.7.

The value of m will have to be estimated from past data, and periodically recalculated.

If for some reason the group size varies, it is better to plot defects per item, and keep recalculating the limits. For further practical details, with worked examples, see Duncan (1974), Grant and Leavenworth (1972), and Huitson and Keen (1965).

With both fraction defective charts (p-charts) and number of defects charts (c-charts), there is the problem of determining the group size and sampling interval, as with normally distributed data; see the references for details.

Exercises 3.6

1. A sample of n items is drawn from a production process every half-hour, and the fraction defective noted. Assume that you have records of, say, 20 groups. What significance test would you apply to determine whether there was evidence that the true fraction defective is homogeneous?

What significance test would you apply if the observation was the total number of defects in the group of n items?

2. If in Figure 3.7 the variable plotted was $\sin^{-1}\sqrt{p}$, where would the action and warning limits be placed? What advantages and disadvantages would this method have over that suggested above?

What transformation would you use for a c-chart?

3.7 Other types of chart for variables

In an important paper Roberts (1966) described five types of control chart for controlling the mean of normally distributed data, with their modifications, and gave a comparison of these based on what is essentially the average run length. Of these five types one is the ordinary Shewhart $\bar{x}$-chart (see section 3.2), and another is the CUSUM chart which will be described in the next chapter; the remaining three types are briefly described below.

(a) Moving average charts

Let $x_1, x_2, \ldots, x_n$ be observations taken at equally spaced intervals, then the moving average of index k is

$$\bar{x}_i^{(k)} = (x_{i-k+1} + x_{i-k+2} + \ldots + x_i)/k.$$

This moving average should really be plotted at the mid-point of the times represented, but it is common practice to plot against the time of the final observation.

The standard error of the moving average is $\sigma/\sqrt{k}$, if σ^2 is the variance of observations. Therefore action limits would be placed at $\pm 3{\cdot}09\sigma/\sqrt{k}$, although any other multiplier could be used instead of 3·09, if desired. Unless moving averages are separated by more than $(k - 1)$ sampling points, the moving averages are correlated, so that it is not possible to use any warning lines as in the ordinary Shewhart $\bar{x}$-chart. Nothing clear can be deduced when successive points are near the action limits.

The ARL of the moving average chart was investigated by Roberts (1959) using simulation on a computer. The moving average chart is much better than the Shewhart $\bar{x}$-chart at detecting small changes from the target value, but less quick at detecting large changes, and both of these properties become more extreme as the index k increases. Therefore a moving average chart can usually be improved by using an ordinary $\bar{x}$-chart in addition.

Moving average charts are particularly suitable where it takes some time to produce a single item. Each new observation can be added as it arises. A simple graphical technique for calculating the moving average chart is described by Roberts (1958). The calculations are also illustrated by Grant and Leavenworth (1972).

Some useful bounds on the ARL curve for moving average charts are given by Lai (1974). These bounds are often quite close. One difficulty with moving average charts is a lack of adequate guidance for designing them.

A similar idea to the moving average chart is the moving range chart. (This is not discussed by Roberts, who confines his discussion to charts for controlling the mean.) Ranges based on a moving group of observations are calculated and plotted

$$\begin{aligned} r_i = & \max(x_{i-k+1}, \ldots, x_i) \\ & -\min(x_{i-k+1}, \ldots, x_i) \end{aligned}$$

against time. Action limits are constructed in the usual way, but warning limits cannot be used because neighbouring moving ranges are highly correlated. The calculation of a moving range chart is illustrated by Grant and Leavenworth (1972). The ARL properties do not appear to have been studied.

(b) *Geometric moving average chart*

The geometric moving average is calculated using the formula

$$z_i = (1 - r)z_{i-1} + rx_i, \tag{3.16}$$

where $z_0 = \mu_0$, the target value. This can be written

$$z_i = (1 - r)^i\mu_0 + r\sum_{s=1}^{i} x_s(1 - r)^{i-s}. \tag{3.17}$$

Therefore the variance of z_i is

$$V(z_i) = \sigma^2 r^2 \sum_{0}^{i-1} (1 - r)^{2s}$$
$$= \sigma^2 r^2 \{1 - (1 - r)^{2i}\}/\{1 - (1 - r)^2\}$$
$$= \sigma^2 r \{1 - (1 - r)^{2i}\}/(2 - r). \qquad (3.18)$$

This variance very quickly approaches its asymptotic value, which can be used as an approximation except at small i,

$$V(z_i) \simeq \sigma^2 r/(2 - r). \qquad (3.19)$$

The procedure is therefore to plot z_i against i, and insert action limits at, say, $\pm 3{\cdot}09$ times the standard error of z_i, that is at

$$\mu_0 \pm 3{\cdot}09\sigma \sqrt{\left(\frac{r}{2 - r}\right)}$$

(For small i, the square root of (3.18) must be used for the standard error.) Warning limits cannot be used.

Bather (1963) showed that the geometric moving average was optimum for a first-order autoregressive process. The ARL of the geometric moving average chart was examined by Roberts (1959) using computer simulation. The ARL properties of the chart are similar to those of the moving average chart, and again it is useful to keep an $\bar{x}$-chart in addition, for quick detection of large changes in the mean.

The geometric moving average chart was suggested by J. W. Tukey, and Roberts (1959) described the procedure and gave a simple graphical method of plotting the points.

4. Cumulative sum charts

4.1 Introduction

The idea of cumulative plotting follows on naturally from the techniques described in the last chapter. We assume that we have inspection of a continuous process, and groups of n items are sampled at a series of equally spaced time intervals. The aim is to keep the process in statistical control, and detect changes in the mean or variance of the observations. The ordinary Shewhart $\bar{x}$-chart only takes the current group of observations into account, and modifications such as warning limits bring the previous group or two into consideration. The geometric moving average and straight moving average charts extend the principle of detecting changes by using previous observations along with the current point. In cumulative sum charts, changes in the mean are detected by keeping a cumulative total of deviations from a reference value. The following artificial example illustrates the basic idea.

The 50 numbers in Table 4.1 are sampled from a normal distribution with mean five and variance unity. If the target value is five, the CUSUM chart for these data is shown as the full line curve in Figure 4.1. The calculations are illustrated in Table 4.2.

Table 4.1. *Observations for CUSUM chart example*

3·80	3·91	5·51	4·37	5·01
5·10	4·07	5·42	4·19	6·33
5·72	4·89	5·77	4·49	5·65
3·70	5·62	5·33	6·24	4·71
3·30	3·98	5·23	5·75	5·34
5·66	4·67	5·57	5·55	4·19
5·33	5·10	5·14	5·32	4·38
5·12	6·08	6·55	3·96	4·95
5·80	4·74	3·33	6·06	5·60
4·44	4·74	4·78	6·46	3·45

Table 4.2 *Calculation of a CUSUM chart for Table 4.1 data*

Observation	*Observation – target value*	*Cumulative sum of column 2*
3·80	−1·20	−1·20
3·91	−1·09	−2·29
5·51	+0·51	−1·78
4·37	−0·63	−2·41
5·01	+0·01	−2·40
5·10	+0·10	−2·30

For the second 25 observations, 0·25 was added to each and the CUSUM chart replotted; this curve is shown as the dotted line in Figure 4.1. While the mean of the observations is equal to the target or reference value, the cumulative sum fluctuates about zero, but as

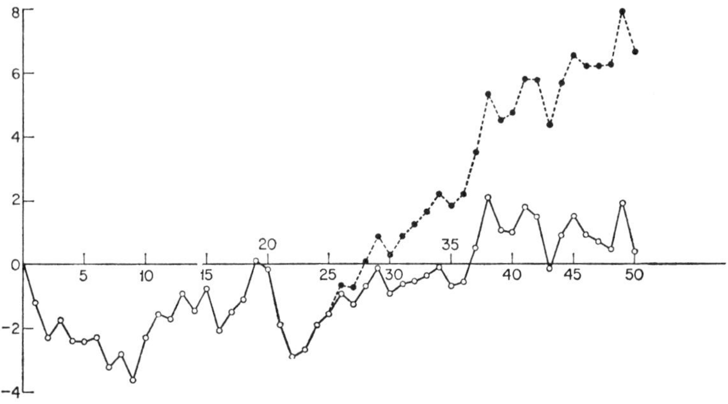

Figure 4.1. *CUSUM chart of Table 4.1 data.*

soon as the mean differs from reference, the cumulative sum begins to increase or decrease.

Let the individual observations be $x_i i = 1, 2, \ldots$, with $E(x_i) = \mu$, and a reference value of m. The quantity plotted is $\sum (x_i - m)$, and

$$E\left\{\sum_{1}^{r} (x_i - m)\right\} = r(\mu - m) \tag{4.1}$$

Therefore the cumulative sum has a slope of $(\mu - m)$ when plotted against r, and CUSUM charts are to be interpreted by the average slope of the graph. If the mean μ is equal to the reference value m, the graph will have very little evidence of a slope; the slight inclination of the full-line graph in Figure 4.1 is due to random variation. As soon as the mean μ differs from m, the graph has an average inclination of $(\mu - m)$; this is shown very clearly by the dotted-line graph in Figure 4.1. The positions on a CUSUM chart at which the graph has a change of inclination indicate the position of possible changes in the mean.

Since on a CUSUM chart it is inclinations which are important, the choice of the scales on the axes should be made with care. A little practice shows that a convenient scale is to choose the distance representing one unit on the horizontal scale to represent 2σ units in the vertical direction, where σ^2 is variance of the short-term variability of the series. (See Ewan, 1963, p. 17, for a discussion of the scale factor for use on CUSUM charts.)

The choice of the reference value must also be made with some care, and it is not satisfactory to use the target value of the process. If the reference value is not equal to the current mean, the graph will slope up or down. This could lead to the graph running off the edge of the paper, and much more seriously, to *reduced* sensitivity to changes in the direction of the slope. A CUSUM chart is most sensitive to changes *from the reference value.*

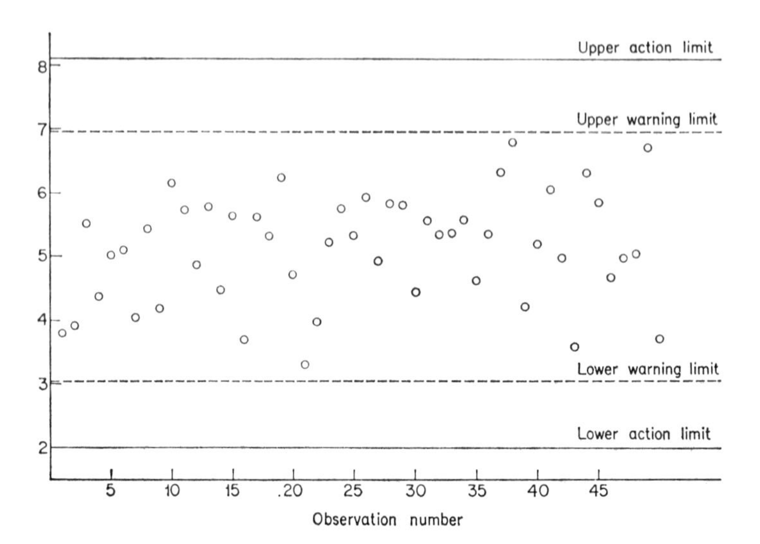

Figure 4.2. *An $\bar{x}$-chart.*

The value of CUSUM charts in comparison with $\bar{x}$-charts can be seen by comparing Figure 4.1 with Figure 4.2. In Figure 4.2, an $\bar{x}$-chart has been plotted for Table 4.1 data, but with the second 25 observations having 0·25 added. We see that whereas the $\bar{x}$-chart gives no indication of a change in mean, the CUSUM chart shows this clearly. We can expect that small deviations from the reference value will show up more quickly (on average) with a CUSUM chart than with a Shewhart $\bar{x}$-chart. In contrast, large deviations from the reference value are likely to show up more quickly with an $\bar{x}$-chart since it takes at least two or three observations to see an inclination on a CUSUM chart. Later in the chapter this comparison will be made more precise. Further examples of CUSUM charts and another comparison with Shewhart $\bar{x}$-charts are given in Woodward and Goldsmith (1964, pp. 6–9).

Before proceeding to a more thorough description and analysis of CUSUM techniques, it may be useful to list some applications:

(*i*) To detect whether a change in mean level has occurred in a continuous process.

(*ii*) To detect whether a change in mean level has occurred in a continuous process, and estimate the amount of the change.

(*iii*) To sort continuous production into categories such as effective or defective.

(*iv*) To search through a set of past data to see when changes in mean level occurred.

(*v*) To indicate changes and trends in sales figures, and form part of a short-term sales forecasting system.

The details of these and other applications are given in Woodward and Goldsmith (1964). In this chapter we are merely concerned with the main principles.

Exercises 4.1

1. Obtain some random normal deviates and repeat the calculations leading to Figures 4.1 and 4.2. Also try adding 0·15 and 0·40 to the second 25 deviates.

2. Plot the data given below on a CUSUM chart, and, remembering that it is changes of inclination which indicate changes in the mean, try and interpret your graph. How will you decide upon a suitable reference value to use?

The data are from a single stage continuous chemical process in which raw materials A and B are reacted together to form a product

C. The reaction is exothermic and water cooling is used to control the reaction temperature to 160°C. The raw materials A and B are both delivered by tanker from which they are run into small stock tanks. There are two stock tanks for each raw material and they are filled and emptied alternately. A full stock tank contains about two weeks' supply of raw material, and each stock tank full is termed a batch. The plant occasionally has to be shut down for cleaning.

The product from the plant is sampled and analysed every shift, and the last 150 observations are tabled below together with an indication of when new batches of raw material were started and when other plant upsets occurred.

Calculate the cumulative sum chart of the data and thence deduce which factors appear to affect the plant efficiency

150 consecutive observations of efficiency and associated plant events

Sample	*Efficiency* (%)	*Comment*
1	46·0	
2	45·5	
3	46·5	
4	44·6	
5	47·3	
6	46·0	
7	45·9	
8	44·5	
9	46·2	
10	44·7	
11	43·9	
12	45·5	
13	45·0	
14	45·5	
15	45·6	
16	43·9	
17	46·2	
18	43·6	
19	45·3	
20	45·9	
21	45·5	
22	44·8	New batch of B
23	45·4	
24	46·7	

Sample	*Efficiency* (%)	*Comment*
25	44·2	
26	44·5	
27	43·8	
28	44·6	
29	45·0	
30	46·0	
31	45·4	
32	44·7	
33	45·3	
34	45·2	
35	45·3	
36	46·1	
37	44·8	
38	46·2	
39	44·4	
40	47·3	
41	46·9	
42	47·1	New batch of A
43	45·5	
44	47·0	
45	45·4	
46	44·9	
47	47·0	
48	46·5	
49	47·0	
50	46·2	
51	44·7	
52	44·4	
53	47·0	Plant shut-down
54	39·9	
55	42·3	
56	46·5	
57	44·1	
58	44·1	
59	46·3	
60	47·5	
61	45·4	
62	46·6	
63	44·1	
64	45·6	
65	45·5	
66	47·5	New batch of B
67	47·0	
68	47·8	

Sample	*Efficiency* (%)	*Comment*
69	44·1	
70	45·1	
71	46·4	
72	48·1	
73	46·3	
74	45·5	
75	45·0	
76	46·7	
77	45·2	
78	46·6	
79	45·0	New batch of A
80	43·0	
81	44·4	
82	45·5	
83	41·9	
84	44·4	
85	45·0	
86	45·0	
87	42·3	
88	43·2	
89	45·3	
90	45·9	
91	44·3	
92	42·9	
93	43·5	
94	43·5	
95	43·4	
96	44·0	
97	43·5	
98	43·8	
99	45·5	
100	45·5	New batch of B
101	43·4	
102	43·9	
103	45·0	
104	43·7	
105	42·9	
106	43·9	
107	43·9	
108	43·9	
109	44·5	
110	44·3	
111	44·4	
112	45·8	

Sample	*Efficiency* (%)	*Comment*
113	44·8	
114	44·1	
115	43·9	
116	44·8	
117	44·9	
118	44·7	New batch of A
119	46·5	
120	45·7	
121	43·7	
122	46·6	
123	45·7	
124	44·8	
125	44·9	
126	46·1	Blockage in cooling water line. High temperatures. (samples 126–134)
127	46·7	
128	44·8	
129	45·6	
130	44·8	
131	45·9	
132	46·7	
133	45·5	
134	44·1	
135	45·1	
136	46·4	
137	45·7	
138	44·8	New batch of B
139	45·2	
140	45·9	
141	45·2	
142	45·3	
143	46·9	
144	46·0	
145	44·7	
146	45·4	
147	45·8	
148	47·1	
149	45·3	
150	44·6	

4.2 Decision rules

(a) The V-mask

Suppose that we are going to use a CUSUM chart to detect when changes in the mean level have occurred; we could then just plot out the chart and look at it to see when corrective action is needed, as indicated in section 4.1. Clearly this subjective approach is not very satisfactory and there is need of some kind of decision rule to indicate when corrective action should be initiated. Two methods are in common use. The methods are equivalent, but each has various advantages.

Barnard (1959) suggested that the V-mask be used. A V-shaped mask is superimposed on the CUSUM chart, the vertex pointing horizontally forwards, and set at a distance d ahead of the most recent point, as shown in Figure 4.3. The angle between the obliques and the horizontal is denoted by θ. If all the previously plotted points fall within the V, the process is assumed to be in statistical control. If some of the points cross one of the arms of the V, a search for assignable causes of variation is initiated. This is illustrated in Figure 4.3.

The properties of the V-mask depend on the choice of d and θ. In his original presentation, Professor Barnard suggested that this choice could be made empirically by cutting a variety of masks and trying them out on past data. However, Goldsmith and Whitfield (1961) evaluated the ARL-curves for a set of V-masks when the observations are independently and normally distributed, by using Monte-Carlo methods. These ARL-curves can be used in choosing a d and θ to use in any particular case.

Johnson (1961) considered the CUSUM chart with a V-mask as (approximately) the operation of two sequential probability ratio tests in reverse order, and in this way he obtained some approximate theoretical results. Johnson also provided some comparisons of the CUSUM chart with the Shewhart chart. The form of these results, and the method of using the ARL curves will be discussed later.

(b) Decision interval schemes

The main differences between this method of operating a decision rule and the one just described are that a graph is not needed, and that the decision rule can be used as either a one-sided or a two-sided test. We describe the one-sided test first.

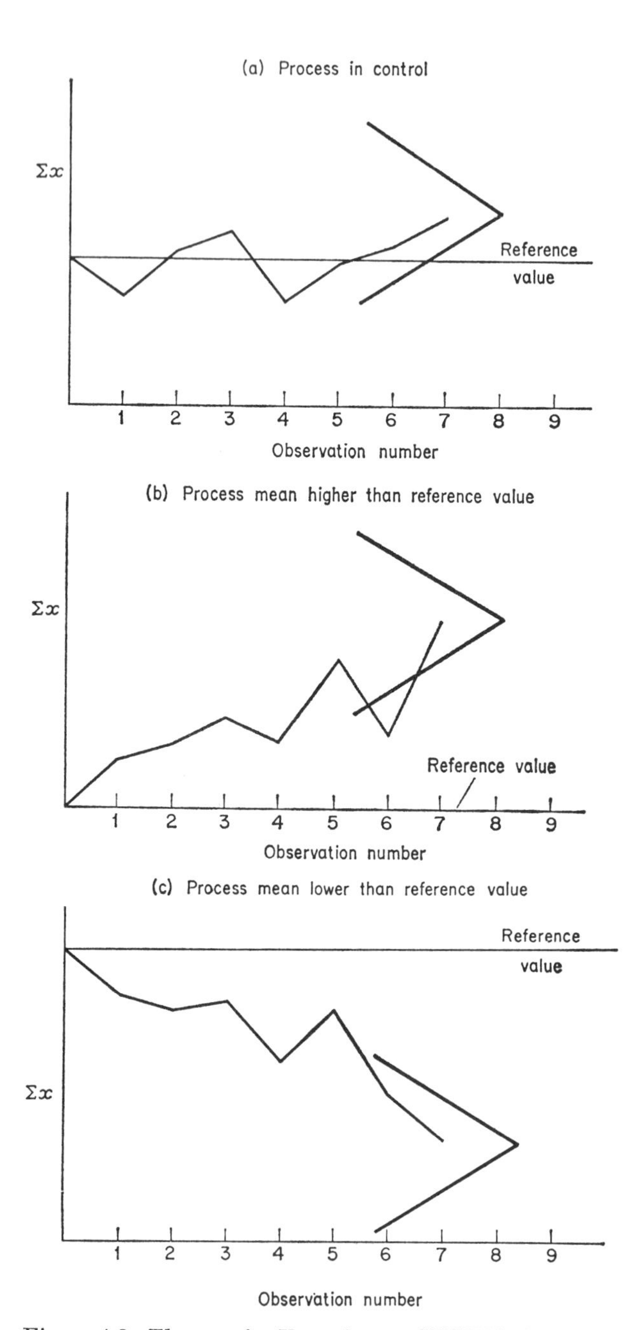

Figure 4.3. *The use of a V-mask on a CUSUM chart.* (*a*) *Process in control,* (*b*) *process mean higher than reference value,* (*c*) *process mean lower than reference value.*

Suppose the reference value for the process is m, and that we wish to guard against the mean level of the observations increasing. Choose a new reference value, k, roughly midway between m and the quality which it is desired to reject. While all observations are less than k, the process is assumed to be in control, and no chart is plotted. As soon as a result exceeds k, a CUSUM chart is started, using k as a reference value. If the chart reverts to zero, the process is assumed to be in control, and no action is required, but if the chart

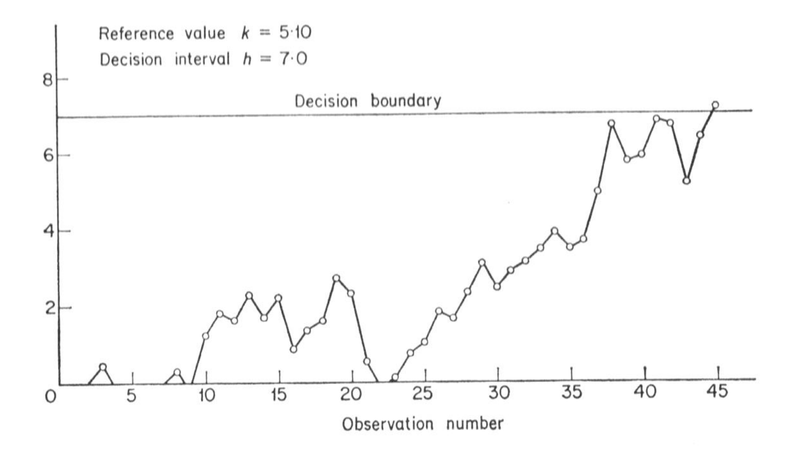

Figure 4.4. *Decision interval scheme for Table 4.1 data.*

exceeds a value h, some corrective action or a search for assignable causes of variation is initiated. The calculations and plotting are illustrated in Table 4.3 and Figure 4.4.

The ARL curves for a set of single-sided decision interval schemes have been derived by Ewan and Kemp (1960), assuming the observations to be independently and normally distributed. The results are briefly described in the next section, and these can be used to choose the parameters h and k in practical cases. The theory is discussed in section 4.4.

For a two-sided decision rule, two single-sided schemes are run concurrently, one for increases and the other for decreases in mean level. The ARL of the combined decision rule is obtained from those of the one-sided rules by using the following formula, which will be

Table 4.3 *Illustration of calculations for decision interval scheme*

Set 1. $k = 5{\cdot}10$, $h = 7{\cdot}0$.

observation	$(x - k)$ *when required*	*control*
3·80	—	—
3·91	—	—
5·51	0·41	0·41
4·37	−0·73	*−0·32*
5·01	—	—
5·10	—	—
4·07	—	—
5·42	0·32	0·32
4·19	−0·91	*−0·59*
6·33	1·23	1·23
5·72	0·62	1·85
4·89	−0·21	1·64
5·77	0·67	2·31
4·49	−0·61	1·70
5·65	0·55	2·25
.	.	.
.	.	.

Set 2. $k = 5{\cdot}75$, $h = 8{\cdot}10$

Observation	$(x = k)$ *when* $x > k$	*Control*
3·80	—	—
3·91	—	—
5·51	—	—
4·37	—	—
5·01	—	—
5·10	—	—
4·07	—	—
5·42	—	—
4·19	—	—
6·33	0·58	0·58
5·72	−0·03	0·55
4·89	−0·86	*−0·31*
5·77	0·02	0·02
4·49	−1·26	*−1·24*
5·65	—	—
3·70	—	—

derived in section 4.4, and which is subject to a condition given there,

$$\frac{1}{(\mathrm{ARL})_{\text{combined}}} = \frac{1}{(\mathrm{ARL})_{\text{upper}}} + \frac{1}{(\mathrm{ARL})_{\text{lower}}}. \tag{4.2}$$

It is important to notice that with this two-sided scheme, accumulations can be running on both sides concurrently.

(c) *Eqivalence of the two decision rules*

Let us assume that groups of n observations are taken at equally spaced time intervals, and let the means of the groups of observations be x_1, $i = 1, 2, \ldots$, each having a variance σ^2/n. For simplicity put the reference value $m = 0$. We shall compare a two-sided decision interval scheme having parameters h and k, with a V-mask scheme having parameters d and θ, in which a unit on the horizontal scale is equal to two standard errors on the vertical scale, or $2\sigma/\sqrt{n}$.

The decision interval schemes will never indicate a need for corrective action if $| x_i | < k$, however many observations there are. The equivalent result for a V-mask scheme is that the increments to the CUSUM chart should not cause a slope more extreme than those of the V-mask.

$$\frac{| x_i |}{2\sigma/\sqrt{n}} < \tan \theta.$$

Hence if we put

$$k = \frac{2\sigma}{\sqrt{n}} \tan \theta,$$

the restrictions would be equivalent.

The V-mask will indicate a need for corrective action at a point P if there is any point Q crossing one of the arms, see Figure 4.5. Therefore the summation from A to B satisfies

$$\sum_{\mathrm{A}}^{\mathrm{B}} x_i > \mathrm{PR} = (d \tan \theta + s \tan \theta)2\sigma/\sqrt{n}$$

which can be written

$$\sum_{\mathrm{A}}^{\mathrm{B}} \left(x_i - \frac{2\sigma}{\sqrt{n}} \tan \theta \right) > \frac{2\sigma}{\sqrt{n}} d \tan \theta, \tag{4.3}$$

but P is the first point after Q at which this inequality is satisfied. If we put

$$k = \frac{2\sigma}{\sqrt{n}} \tan \theta \tag{4.4}$$

$$h = \frac{2\sigma}{\sqrt{n}} d \tan \theta \tag{4.5}$$

equation (4.3) defines a decision interval scheme. Therefore the two decision rules are exactly equivalent, and equations (4.4) and (4.5) give the relationships between the two sets of parameters, where k is measured from m.

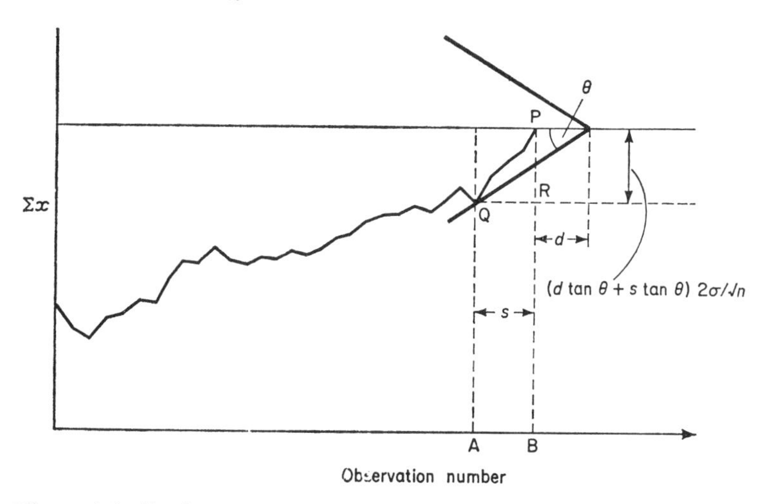

Figure 4.5. *Equivalence of decision rules for a CUSUM chart.*

The value of the decision interval scheme is that for much of the time when production is in control, no score need be kept at all, and even when accumulation is in progress, no chart is needed. However, the CUSUM chart does provide a very good picture of the behaviour of the process, and there may be circumstances when it is desirable to plot this and use a V-mask.

(d) Simplified decision rule

Ewan (1963) described a simplified decision rule which can be used instead of a V-mask. It arises because operators tend to test using a mask only when they see an apparent change.

First the operator estimates visually the apparent point of change and a line AB is drawn giving the path of the chart up to this point. At the estimated point of change a point C is marked at a distance h from this line, and another point D is plotted n points further on, at a distance $(h + nk)$ from the line. (The points C and D are plotted in the

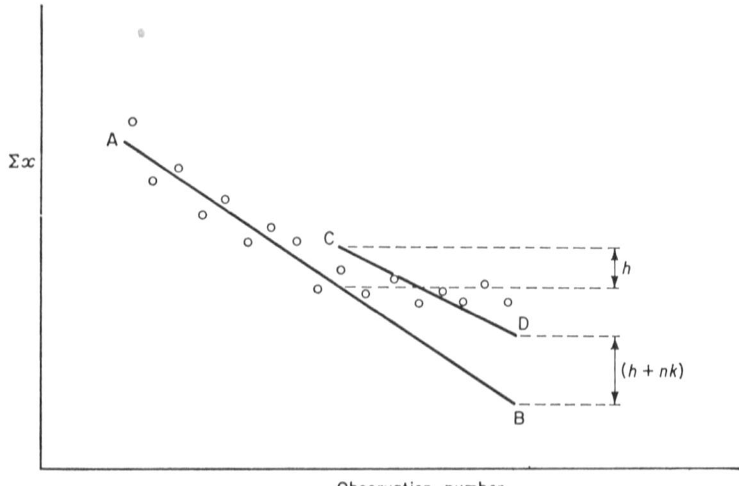

Figure 4.6. *Simplified CUSUM decision rule.*

direction of the apparent change; it is not necessary to wait for n points after an apparent change to plot these points.) If the chart crosses CD a change is declared; see Figure 4.6.

This procedure involves a subjective element in estimating the point of change and drawing the original straight line. However, it has been found very useful in practice, and combines some of the advantages of the two decision rules described above.

(e) *Estimation of σ for CUSUM charts*

We shall assume throughout that there is sufficient back data to form a good estimate of σ for use in setting the scales and the decision rule. If observations are sampled in groups, this estimate is easily obtained from a pooled estimate of within-group variances or from the average of the group ranges.

If observations are taken singly, then the usual estimate is to base the calculation on successive differences,

$$s^2 = \frac{1}{2(n-1)} \sum_{i=1}^{n-1} (x_{i+1} - x_i)^2.$$

This estimate is biased if the series of observations are autocorrelated.

A further complication arises when there are several components of variance present. For example, there may be day-to-day variation, and long-term variation over a day, in addition to the short-term variation which can be estimated by the methods given above. The short-term variance should generally be used in scaling the axes of the CUSUM chart, but knowledge of the other components of variance is required in setting the decision rule. The vital point is to design the chart to be sensitive to the smallest order of change in the mean which is regarded as practically important.

Exercises 4.2

1. For the data of Exercise 4.1.1 calculate a decision interval scheme using $h = 2{\cdot}24$, $k \neq 1{\cdot}12$. Also carry out the calculations for the equivalent V-mask scheme.

4.3 Properties of the decision rules

The ARL-curves of the V-mask have been calculated by Monte-Carlo methods, and Figure 4.7 gives one of these curves, for $d = 2$. Woodward and Goldsmith (1964) also give ARL-curves for $d = 1, 5, 8$, and clearly, as d increases for any given value of tan θ, so does the ARL. These ARL-curves assume that the horizontal unit on the CUSUM chart is $2\sigma/\sqrt{n}$ on the vertical axis, but conversion to other scales is easy.

Let us suppose that the group sample size n is given, and see how these ARL-curves can be used to choose a d and θ. We could define values of the mean at which the quality is acceptable and rejectable; these values are denoted AQL and RQL respectively. We must now choose the ARL's we desire at these two quality levels, and we denote these by L_0 and L_1 respectively. We can now search along the graphs for a pair (d, θ) which approximately satisfies our requirements. (Not all pairs L_0, L_1 can lead to a scheme.) If the group size is not specified in advance, the choice is more difficult since the scale of Figure 4.7 is not fixed. This method of choosing a CUSUM chart is very similar to the method set out in section 2.4(a) for choosing a sampling inspection plan.

The choice of a decision interval scheme can be made in similar manner as for the V-mask scheme, by fixing the ARL's desired at the AQL and the RQL. However, the parameter k in a decision interval scheme has an interpretation, and represents a quality roughly midway between the AQL and RQL. Ewan and Kemp suggest choosing k exactly half-way between the AQL and the RQL, and if the group

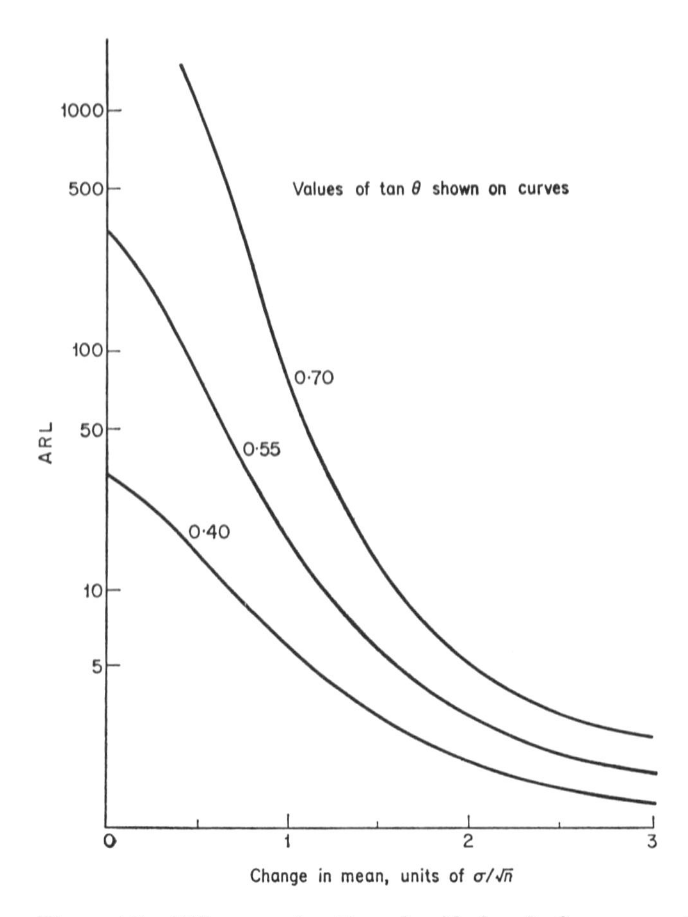

Figure 4.7. *ARL-curve of a V-mask with d = 2. Curves are shown for three values of tan θ. Based on Woodward and Goldsmith 'Cumulative Sum Techniques', Fig. 5b (Oliver and Boyd, 1964), by permission of Imperial Chemical Industries Ltd.*

size is fixed, only one ARL, at either the AQL or the RQL, can be chosen arbitrarily. If the ARL is fixed at the AQL, the resulting ARL at the RQL may turn out to be unacceptable, in which case k will have to be modified (or equivalently, the AQL and RQL). If the group size n is not fixed in advance, k can be chosen midway between the AQL and the RQL, and the ARL's can still be specified arbitrarily. A nomogram to design decision interval schemes is given in Appendix II, and an illustrative example is given in section 4.5.

There seems no clear reason why k equal to the average of the AQL and RQL should be best, especially since the AQL and the RQL are themselves often chosen rather arbitrarily. However, this rule does give a very useful starting-point for finding a plan.

Very little has been said about the choice of group size and sampling interval, and an analysis parallel to that given in section 3.4 is appropriate. However, the qualitative conclusions are liable to be the same. There is also the possibility that the sampling interval could be reduced when quality deteriorates; this type of plan is discussed in the next chapter.

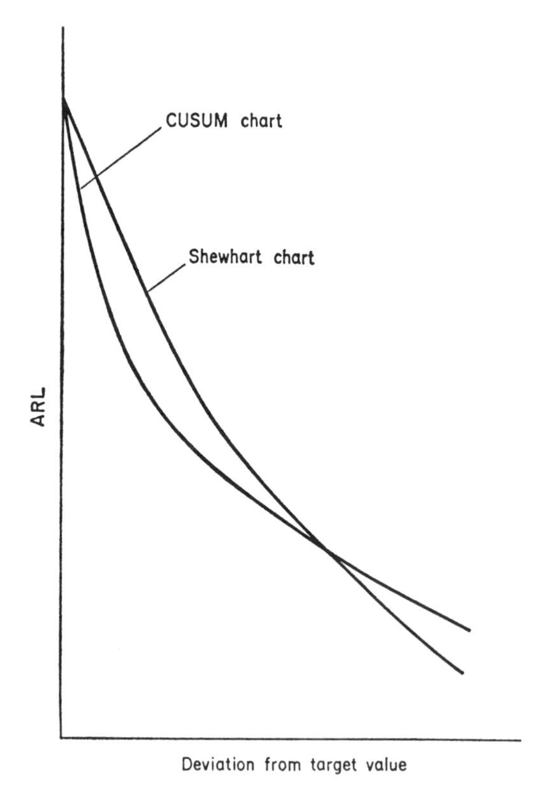

Figure 4.8. *ARL-curve for a CUSUM chart and a Shewart $\bar{x}$-chart.*

One difficulty about the results on ARL-curves described in this section is that they are derived from the assumption that observations are independently and normally distributed. In practice, observations are sometimes correlated, and may not be normal.

Some of the effects of such departures are described briefly by Woodward and Goldsmith (1964, pp. 17–18) and Bissell (1969). In general, the ARL results are likely to be sufficiently robust to these departures to choose a suitable plan for CUSUM charts on a normal variable.

An interesting question now arises; to see how the ARL-curves of Shewhart $\bar{x}$-charts and CUSUM charts compare. This problem was discussed by Roberts (1966), who also compared a number of other schemes. Roberts's general result was that there was surprisingly little difference between the methods. As far as a comparison of the CUSUM chart and the Shewhart $\bar{x}$-chart is concerned. Figure 4.8 illustrates the form of the results. The CUSUM chart detects small changes more quickly than the $\bar{x}$-chart, but the reverse is true for large changes; the reason for this was indicated in section 4.1.

A comparison of the $\bar{x}$-chart and CUSUM chart was also given by Johnson (1961), using an approximate argument. The action lines for Johnson's $\bar{x}$-chart were placed at varying positions, including the conventional position. As the action lines are brought in, the $\bar{x}$-chart gradually becomes superior to the CUSUM chart; see the next section for details.

4.4 Theory of the decision interval scheme

(a) Exact theory

The decision interval scheme illustrated in Figure 4.4 can be considered as a sequence of sequential probability ratio tests; (for a description of and references to this see Wetherill (1975)). Let the variable z denote the distance from the lower boundary of the scheme, so that the two boundaries are at $z = 0$ and $z = h$. A single test is defined as a path starting at a value z satisfying $0 \leq z < h$, and ending at the upper or lower boundary; in the degenerate case the path may be one point only. For such a test let

$P(z)$ = probability that a test starting at z ends at $z \leqslant 0$.

$N(z)$ = average sample number of a test starting at z.

The decision interval scheme is a series of such tests, and terminates with the first test to cross the upper boundary. Let $L(z)$ denote the ARL of a decision interval scheme in which the first test starts at the point z, but all subsequent tests start at the lower boundary.

So far in this chapter we have only discussed CUSUM schemes in which the observations are normally distributed, but to be general, we denote the probability density function of the observations by $f(x)$, and cumulative distribution by $F(x)$. The CUSUM scheme proceeds by observing x, and if the current score is z, the new score is

$$z + x - k \qquad \text{if } x \geq k - z$$

or

$$0 \qquad \text{if } x \leq k - z.$$

We begin by considering a single test starting from a score z, and obtain a formula for $P(z)$. If one observation is taken, there are three possibilities:

	Observation	*New score*	*Outcome*
(*i*)	$x \leqslant k - z$	0	test ends at lower boundary
(*ii*)	$k - z \leqslant x \leqslant h + k - z$	$z + x - k$	test in progress
(*iii*)	$x \geqslant h + k - z$	h	test ends at upper boundary

The probability of the first event is $F(k - z)$. If the second event happens, there is a further probability $P(y)$, for every $y = z + x - k$, $0 < y < h$, of ending at the lower boundary. The last event is irrelevant to $P(z)$. Therefore we have the equation

$$P(z) = F(k - z) + \int_0^h P(y)\, f(y + k - z)\, dy. \tag{4.6}$$

In a similar way we can obtain the equations

$$N(z) = 1 + \int_0^h N(y)\, f(y + k - z)\, dy, \tag{4.7}$$

and

$$L(z) = 1 + L(0)\, F(k - z) + \int_0^h L(y)\, f(y + k - z)\, dy. \tag{4.8}$$

Equations (4.6) and (4.7) have been described by Page (1954), and Page (1954) and Kemp (1958) gave numerical methods for solving them.

The ARL of the decision interval scheme is $L(0)$, and once $P(0)$ and $N(0)$ are obtained by solving (4.6) and (4.7), $L(0)$ can be obtained from the formula

$$L(0) = N(0)/\{1 - P(0)\} \tag{4.9}$$

instead of by solving (4.8) directly. This formula can be derived as follows. In a decision interval scheme the number of sequential tests has the geometric distribution

$$\{P(0)\}^{(s-1)}\{1 - P(0)\}, \qquad s = 1, 2, \ldots.$$

Thus on average there are $\{1 - P(0)\}^{-1}$ sequential tests in a single run of a decision interval scheme of which just one terminates on the upper boundary. If $N(0)^u$, $N(0)^l$ are the average sample numbers of sequential tests terminating on the upper and lower boundary respectively, the ARL of the decision interval scheme is

$$
\begin{aligned}
L(0) &= N(0)^u + \left\{\frac{1}{1 - P(0)} - 1\right\} N(0)^l \\
&= \frac{1}{1 - P(0)} \left\{(1 - P(0))\, N(0)^u + P(0)\, N(0)^l\right\} \\
&= \frac{N(0)}{1 - P(0)}.
\end{aligned}
$$

Now the ARL is the expectation of the distribution of run length, and it is very useful to have a formula for it. However, further information about the run length distribution can easily be obtained. Let $p(n, z)$ = probability that a test starting at z has run length n, then by following an argument similar to that leading to (4.6) we have

$$
p(n, z) = p(n - 1, 0)\, F(k - z) + \int_0^h p(n - 1, y) f(y + k - z)\, \mathrm{d}y. \tag{4.10}
$$

Denote the moment generating function of the run length distribution by $\phi(z, t)$

$$
\phi(z, t) = \sum_1^\infty p(n, z)\, \mathrm{e}^{nt},
$$

then from (4.10) we have

$$
\begin{aligned}
\mathrm{e}^{-t}\phi(z, t) = 1 - F(h + k - z) &+ \phi(0, t)\, F(k - z) \\
&+ \int_0^h \phi(y, t) f(y + k - z)\, \mathrm{d}y. \qquad (4.11)
\end{aligned}
$$

By successively differentiating (4.11) and putting $t = 0$ we can obtain integral equations for the moments of the run length distribution. Ewan and Kemp (1960) also obtained an approximation for the variance of the run length distribution,

$$
V(n) \simeq L^2(0) + V(N)/\{1 - P(0)\}, \tag{4.12}
$$

where $V(N)$ is the variance of the sample number of a single sequential test, and the approximation is valid when $P(0)$ is close to unity. The authors also conjectured that a close approximation to the run length distribution is

$$
p(n, 0) \simeq \frac{1}{L(0)} \exp\left\{-\frac{(n - 1)}{L(0)}\right\}. \tag{4.13}
$$

Throughout this theory, we have assumed that the observations x are continuous, but the methods used can be followed through in the discrete case also. Ewan and Kemp (1960) gave values of the ARL for the case when the observations have a Poisson distribution, as well as for the normal distribution case.

(b) Johnson's approximate approach

Johnson (1961) gave an approximate approach for a CUSUM chart with a V-mask, which arrives at some remarkably simple answers.

We first reverse a CUSUM chart, and look at it as if it were proceeding backwards. Figure 4.9 shows approximately how Figure

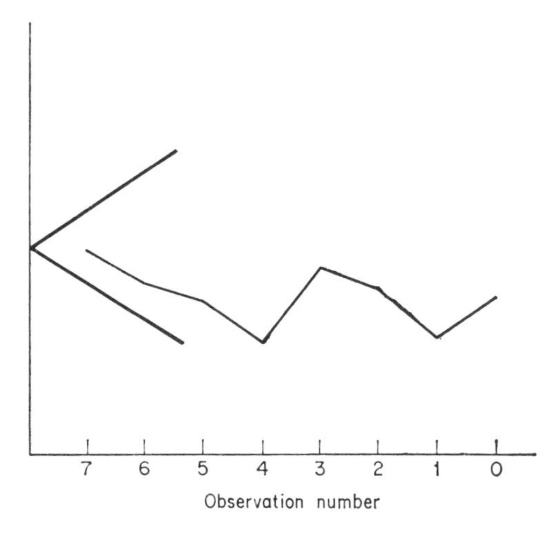

Figure 4.9. *Johnson's approach to V-mask theory.*

4.3(*a*) would be reversed. The method is now to regard the outer arms of the V-mask as boundaries of a test of three simple hypotheses using the sequential probability ratio test (see Wetherill, (1975) and references).

Suppose we have three hypotheses, that observations are independently and normally distributed with distributions as follows:

$$H_{-1}: N(-\delta\sigma, \sigma^2); \quad H_0: N(0, \sigma^2); \quad H_1: N(\delta\sigma, \sigma^2).$$

Suppose, further, that we want a probability $(1 - 2\alpha_0)$ of accepting H_0 if it is true, and a probability $(1 - \alpha_1)$ of accepting H_1 or H_{-1} if they are true, then the boundaries for the sequential probability ratio test of these hypotheses are as illustrated in Figure 4.10.

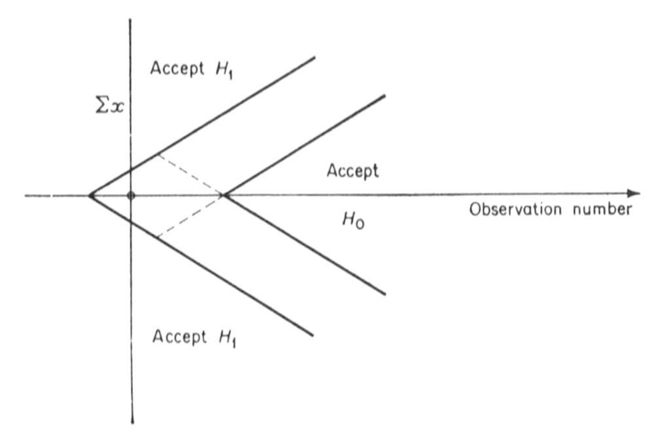

Figure 4.10. *Sequential tests of three hypotheses.*

The outer boundaries are

$$\sum_{i=1}^{n} x_i = \left[\frac{1}{\delta}\log_e\{(1-\alpha_1)/\alpha_0\} + \tfrac{1}{2}\delta n\right]\sigma \tag{4.14}$$

and

$$\sum_{i=1}^{n} x_i = -\left[\frac{1}{\delta}\log_e\{(1-\alpha_1)/\alpha_0\} + \tfrac{1}{2}\delta n\right]\sigma. \tag{4.15}$$

Now the outer boundaries of the V-mask in Figure 4.5, using the current point as origin, are

$$\sum x = 2\sigma \tan\theta(n+d) \tag{4.16}$$

and

$$\sum x = -2\sigma \tan\theta(n+d) \tag{4.17}$$

where 2σ is the scale factor of the CUSUM chart. It follows that if we identify (4.14) with (4.16) and (4.15) with (4.17), we shall have a V-mask in which, approximately, the probability of a path crossing an outer boundary is $2\alpha_0$, when the process is in control. By identifying these pairs of equations we obtain

$$\tan\theta = \delta/4 \tag{4.18}$$

$$d = 2\log_e\{(1-\alpha_1)/\alpha_0\}/\delta^2. \tag{4.19}$$

Unfortunately α_1 in this last equation is difficult to interpret, since there is no 'accept H_0' boundary on Figure 4.5. However, since α_1 is usually small, we have

$$d \simeq -2\log_e(\alpha_0)/\delta^2. \tag{4.20}$$

These results can be used in the following way. First decide on the least change in the mean which it is desired to detect with reasonable certainty; let the standardized value, standardized by σ, be δ. We must now decide on the greatest tolerable probability, $2\alpha_0$, of false indications of lack of control; values near 0·002 are traditional for this in control chart work. Use of (4.18) and (4.20) now give θ and d corresponding to this pair of (δ, α_0). The properties of the selected (d, θ) can be checked from tabulated ARL-curves, and modified if they are not satisfactory.

Johnson points out that this theory throws some further light on CUSUM charts. Since CUSUM charts are like a two-sided SPRT without a middle boundary, and there is no decision to 'accept H_0', a path which would have been terminated on an SPRT could go on and cross one of the decision boundaries. Therefore paths which cross the decision boundaries a long way from the vertex should be regarded with suspicion.

(c) *Proof of equation* (4.2)

Let $x_1, x_2, \ldots, x_N$ be observations from which a two-sided decision interval scheme such as that shown in Figure 4.11 is operated.

In Figure 4.11, once a decision boundary is crossed, the chart automatically restarts at zero. It can easily be shown that if, say, the upper decision boundary is crossed, plotting on the lower chart will have terminated at the 'in control' boundary: see Exercise 4.4.4. Therefore this automatic resetting of the scheme has no effect on the plotting.

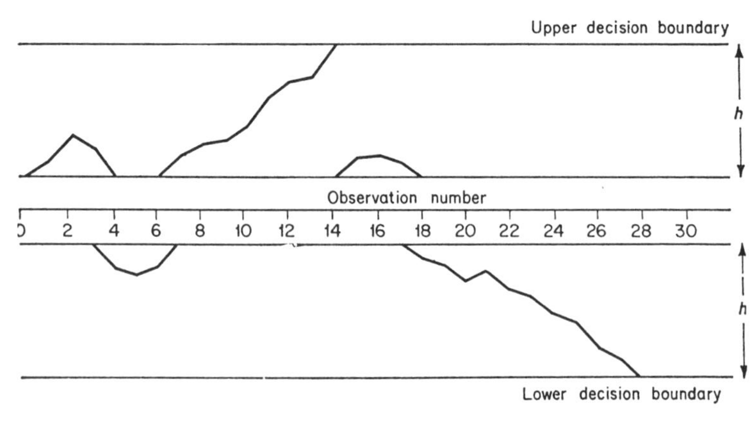

Figure 4.11. *Series of two-sided decision interval schemes.*

Let

x = number of terminations at a decision boundary
x_u = ,, ,, ,, ,, the upper decision boundary
x_l = ,, ,, ,, ,, ,, lower ,, ,,

then clearly $x = x_u + x_l$, and

$$E(x) = E(x_u) + E(x_l).$$

But $$E(x_u) = N/(\text{ARL})_{\text{upper}},$$

and $$E(x_l) = N/(\text{ARL})_{\text{lower}},$$

hence equation (4.2) is obtained.

If the parameters for the upper and lower chart are different, say (k^+, h^+) (k^-, h^-) for the upper and lower charts respectively, then Dobben de Bruyn (1968) showed that (4.2) remains true provided

$$k^+ + k^- \geq | h^+ - h^- |.$$

This condition is necessary and sufficient to establish that, if one decision boundary is reached, plotting on the other chart will have terminated at the 'in control' boundary.

Exercises 4.4

1*. Refer back to the details in Ewan and Kemp (1960) and show how to evaluate the ARL for a CUSUM scheme on variances of normally distributed data.

2*. Study how to set about a thorough investigation into the effects of deviations from assumptions, such as non-normality and serial correlation upon the run length distribution of a decision interval scheme.

3*. All our treatment of CUSUM schemes has assumed that observations are taken in groups of n at equally spaced intervals. Set out a model for examining an optimum choice of group size and sampling interval. (You may be guided by similar work referred to in earlier chapters.)

4*. For the two-sided decision interval scheme discussed in section 4.4(c), let the two reference values be $\pm k$. Show that if plotting on one chart ends at a decision boundary, plotting on the other must have ended at the 'in control' boundary. (See Kemp 1961, p. 151.)

4.5 Use of nomogram to design decision interval schemes

In Appendix II a nomogram produced by Kemp (1962) has been reproduced, and this can be used to derive parameters for operating a decision interval scheme.

Suppose that in some application discussion with staff establishes that a one-sided scheme is required, with

$$\text{AQL} = 11{\cdot}10; \quad \text{RQL} = 12{\cdot}66; \quad \sigma = 2;$$
$$L(\mu = 11{\cdot}10) = 800; \quad \text{L}(\mu = 12{\cdot}66) = 5;$$

where $\text{L}(\mu = 11{\cdot}10)$ denotes the ARL at the AQL, etc. If the reference value is chosen midway between the AQL and RQL this gives

$$k = (11{\cdot}10 + 12{\cdot}66)/2 = 11{\cdot}88.$$

If a ruler is put across the nomogram corresponding to the two ARL's, the other two boundaries are cut at approximately

$$\frac{h\sqrt{n}}{\sigma} = 3{\cdot}32, \quad | \mu - k | \frac{\sqrt{n}}{\sigma} = 0{\cdot}78$$

but since at the RQL,

$$(\mu - k) = 12{\cdot}66 - 11{\cdot}88 = 0{\cdot}78$$

then

$$\sqrt{n}/\sigma = 1, \quad \sqrt{n} = 2$$

and hence the required scheme is

$$h = 3{\cdot}32, \quad k = 11{\cdot}88, \quad n = 4.$$

In order to plot the ARL curve, some further values of the ARL are required. If we consider $\mu = 11{\cdot}38$ or $12{\cdot}38$, then for the same decision interval scheme,

$$| \mu - k | \frac{\sqrt{n}}{\sigma} = 0{\cdot}5, \quad \frac{h\sqrt{n}}{\sigma} = 3{\cdot}32$$

A ruler must now be placed across the nomogram at these values, to read off

$$L(\mu = 11{\cdot}38) = 132, \quad L(\mu = 12{\cdot}38) = 7{\cdot}2$$

A freehand curve can now be drawn by plotting log (ARL) against μ, and this will be sufficiently accurate for most purposes.

In the example used above, the value of n required came out exactly to an integral number. Clearly this will not be the case in general, and it will be necessary to try out, say, the nearest integral value. One of the ARL's will then have to be slightly different from the values specified. Alternative possibilities can then be worked by calculating the new value of $(\mu - k)\sqrt{n}/\sigma$, and fixing one of the ARL's.

If a two-sided decision interval scheme is required, we must use equation (4.2). Suppose we were given an AQL of 11·10, and RQL's of 9·54 and 12·66 with required ARL's $\text{L}(11{\cdot}10) = 400$, $\text{L}(9{\cdot}54) =$

$L(12{\cdot}66) = 5{\cdot}0$. Then at the RQL, the ARL is very nearly that of a one-sided scheme, but application of (4.2) at the AQL leads to $L(\mu = 11{\cdot}10 \mid \text{one-sided}) = 2L(\mu = 11{\cdot}10 \mid \text{two-sided})$. Hence we are looking for one-sided schemes with $L(\mu = 11{\cdot}10) = 800$, $L(\mu = 9{\cdot}54) = L(\mu = 12{\cdot}66) \simeq 5{\cdot}0$, and this was the case discussed above.

Before employing the method given above too widely, it is important to have some appreciation of the effect of deviations from the various assumptions.

Firstly, a value of σ has to be assumed, and it is quite clear that the effect of departures from the assumed value on the ARL can be dramatic; see comments on this in De Bruyn (1968, pp. 44, 45). Overestimation of σ increases the ARL, and underestimation reduces it. Great care must be taken over the choice of σ.

Bissell (1969) studied the effect of skewness of the underlying distribution, and provided a nomogram to assess the effect on the ARL. His general conclusion is intuitively clear from the way in which CUSUM cumulations arise. At the RQL, most of the distribution contributes to the cumulations, and the effect of skewness is very small, but the position is different at the AQL. For positive skewness, the proportion (and mean) of observations contributing to cumulations will increase, while at the same time the proportion (and mean) of observations detracting from cumulations will decrease. The result is that positive skewness can seriously reduce the ARL at the AQL. By similar reasoning, negative skewness increases the ARL at the AQL.

The effect of serial correlation between observations has been studied by Goldsmith and Whitfield (1961) using simulation, and by Johnson and Bagshaw (1974) and Bagshaw and Johnson (1975) by theoretical means. The general conclusion is similar to the case of skewness mentioned. Positive serial correlation tends to reduce the ARL, and negative serial correlation tends to increase the ARL. Again there is little effect at the RQL, but at the AQL the effect can be quite large. Johnson and Bagshaw (1974) say: 'Our primary conclusion is that the Cusum test is not robust with respect to departures from independence. The use of Cusum tests is now widespread, and the presence of serial correlation common so that attention should be drawn to the seriousness of this lack of robustness.'

Exercises 4.5

1. A one-sided decision interval scheme is run with parameters $h = 7{\cdot}6$, $k = 50$, AQL $= 47{\cdot}5$, RQL $= 52{\cdot}5$, $n = 9$, $\sigma = 8$. Sketch the ARL curve.
2. A one-sided decision interval scheme is required where the AQL is 4·60, the RQL is 5·20, and the corresponding ARL's are to be 800 and 4·0 respectively. Samples are taken every 15 minutes, and this is fixed, but the sample size can be varied, and the standard deviation of observations is 0·667. Find suitable values of h, k, and n.
3. If in the previous exercise $\sigma = 0{\cdot}80$, show that there is no value of n which will satisfy the requirements. Examine various possible values of h, k, and n which will approximate to a solution.
4. You are given that the AQL $= 0$, RQL $= 1{\cdot}96$, $\sigma/\sqrt{n} = 1$, then:
   (*a*) if samples of n are taken every hour and $L(\mu = 0) = 250$, find the ARL at the RQL;
   (*b*) if samples of $n/2$ are taken every half-hour and $L(\mu = 0) = 500$, find the ARL at the RQL;
   (*c*) if samples of $n/3$ are taken every 20 minutes and $L(\mu = 0) = 750$, find the ARL at the RQL.

Hence comment on the effect of sampling interval on the ARL properties of CUSUM schemes.

5. Suppose that, for the situation described in Exercise 4.5.2, a two-sided decision interval scheme is required where the AQL is 4·60 with an ARL of 400, and the RQL's are 4·00 and 5·20, with an ARL of 4·0. Find suitable values of h, k, and n, and the equivalent values of d and θ for a V-mask scheme.

4.6 The economic design of CUSUM control charts

Taylor (1968) has given a discussion of the economic design of CUSUM charts when used for the purpose of controlling a process. The model used is similar to the one used by Duncan (1956) for the economic design of $\bar{x}$-charts.

Groups of n observations are taken every h hours from a manufacturing process. The observations are assumed to be normally distributed with a constant variance σ^2, and a mean μ_0 while the process is in control. After a time T the process goes out of control and the mean changes to

$$\mu_1 = \mu_0 \pm \delta\sigma/\sqrt{n}.$$

Therefore δ measures the change in the mean in standard error units;

this assumption has the unfortunate result that we alter the change in the process mean by altering n.

Let M be the number of observations until the process goes out of control, so that

$$Mh < T < (M + 1)h$$

and let F be the number of false out-of-control signals given by the chart in this time. Let S be the number groups of observations taken after the process goes out of control until the chart indicates lack of control. Assume that when the chart gives an out-of-control signal, the process is shut down for an average time τ_s, and if the signal is not a false alarm, a further time τ_r is taken to repair the process. Once the process is repaired, the mean is again μ_0, and the cycle is repeated. These cycles are of average length

$$\tau_r + \tau\{E(F) + 1\} + hE(M + S). \tag{4.21}$$

We now introduce the cost parameters. Denote:

p = profit rate per hour when the process is in control
c = cost ,, ,, ,, ,, ,, ,, ,, out of control.
k_s = cost of a search for lack of control
k_r = cost of repairing the process.

If we assume that the search and repair times are independent of M and T, then the average cost per cycle is

$$k_r + k_sE(F + 1) + cE\{(M + S)h - T\} - pE(T). \tag{4.22}$$

The expected total cost per hour is (4.22) divided by (4.21), but to be of any use, we must evaluate $E(F)$, $E(S)$, and $E(M)$.

Write $E(T) = \mu_T$, and assume this known; then as an approximation take

$$E(M) = \mu_T/h - \tfrac{1}{2}. \tag{4.23}$$

Let the ARL's of the CUSUM chart in control and out of control be $L(0)$ and $L(\delta)$ respectively. The ARL's $L(0)$ and $L(\delta)$ are those obtained in the previous section under the assumption that the chart starts from zero with the given mean (either μ_0 or μ_1). Then we can approximate

$$E(F) = E(M)/L(0) \tag{4.24}$$

and

$$E(S) = L(\delta). \tag{4.25}$$

The last step involves ignoring the effect of head starts which a chart may have when the process goes out of control, and the results of some simulation runs are given to justify this.

If we use (4.23), (4.24) and (4.25) in (4.22) and (4.21), we obtain for the average cost per hour

$$C = \frac{k_r + k_s\{1 + (\mu_T/h - \frac{1}{2})/L(0)\} - p\mu_T + ch\{L(\delta) - \frac{1}{2}\}}{\tau_r + \tau_s\{1 + (\mu_T/h - \frac{1}{2})/L(0)\} + \mu_T + h\{L(\delta) - \frac{1}{2}\}}. \quad (4.26)$$

Now if the V-mask is used on the CUSUM chart, $L(0)$ and $L(\delta)$ will be functions of the parameters d and θ of the mask. We must therefore choose (d, θ, n, h) to minimize (4.26), and this can be done fairly readily on a computer by using the approximate formulae for $L(0)$ and $L(\delta)$ given by Goldsmith and Whitfield (1961).

Taylor simplifies the optimization problem as follows. He argues that a choice of θ such that

$$(\text{scale factor}) \tan \theta = |\mu_1 - \mu_0|/2$$

would be approximately optimal (see 4.18) and he gives some numerical evidence to support this. If, as sometimes happens, n and h are given, the only remaining parameter is d. A simple numerical example is given in the paper.

In any practical case it should be easy to obtain information on the average times μ_T, τ_s, and τ_r. If the process is not shut down during a search, this could be allowed for. The most unrealistic assumption is that concerning the mean of the process, but it would be easy to allow for a distribution of out-of-control means over a finite number of values, provided some information was available on the true distribution of means.

It would be interesting to have an extensive numerical exploration of this optimal solution, in particular to see if certain cost parameters, or some functions of them, are really critical.

4.7 Estimation from CUSUM charts

The methods given in section 4.2 were merely decision rules for deciding when to initiate investigations for assignable causes of variation. That is, they are merely rules to say when there is evidence that the mean has changed, and there is no attempt to estimate the amount of this change. In many practical situations an estimate of the amount of the change is vital in order to give a guide to the extent of any adjustment which may be necessary. We shall deal first with an estimation method for the decision interval scheme, since this is very much simpler than the estimation method for a CUSUM plot with a V-mask.

The crux of the problem is that the mean has probably changed, and at a point which is not known precisely. It is necessary to use the CUSUM chart to dictate how much past data should be used in making an estimate of the new mean.

(a) Estimation method for decision interval scheme

In a decision interval scheme, a series of runs are plotted, some terminating at the 'in control' boundary, and others terminating at the decision boundary. It is therefore reasonable to take as an estimate of the new mean, the average of the observations which compose the run terminating at the decision boundary. This estimate is biased, but no correction has been suggested.

Suppose a two-sided decision interval scheme is operated about a central reference value of zero with reference values of $\pm k$ for the two schemes, then the estimation procedure just suggested has the property that it is impossible to obtain an estimate within the range $(-k, +k)$. However, if the estimate is required for some adjustment, this inert region will not matter since no decision to apply an estimate in this region could be reached.

No further properties of this estimation procedure have been investigated. In particular, if estimation were the main aim and a decision rule a secondary consideration, a rather different choice of h and k ought probably to be used.

(b) Estimation procedure for the V-mask scheme

A method of estimation for use with a straight CUSUM chart and V-mask scheme, must be a means of picking out sections of the chart

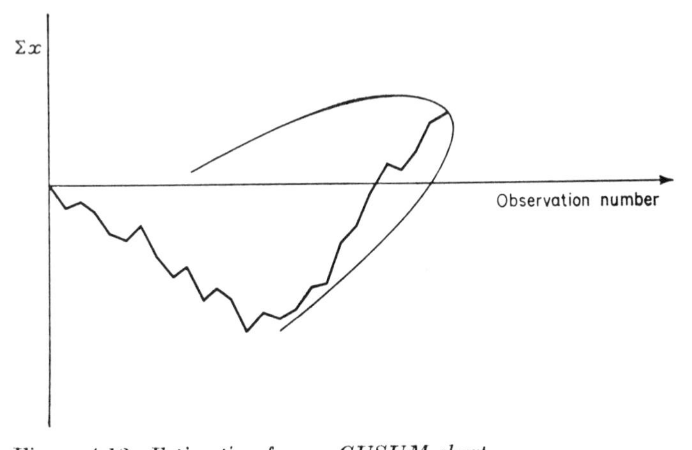

Figure 4.12. *Estimation from a CUSUM chart.*

which are of fairly constant inclination. Barnard (1959) suggested that a parabolic mask be used for this purpose.

Briefly, the method is to place the vertex of the parabola over the current point, and then rotate the parabola so as to include as many of the past points as possible. This is illustrated in Figure 4.12. The direction of the axis of the parabola is taken to estimate the mean. Clearly, the parabola will tend to include one or two points too many, but the bias introduced by this will be small. One counter-measure against this is to include as many points as possible, but rotate the parabola away from the point which is just not included until it passes through one of the points which are included.

Very little is known about the properties of this method, and although a parabola is obviously roughly the right shape, there is no certainty that it is the best. Various practical points concerning the use of the technique are discussed by Woodward and Goldsmith.

Lucas (1973) has suggested that a parabolic mask be used instead of a V-mask more generally. However, the schemes are more difficult to apply, and we have not a ready means of designing such masks or of determining their properties.

4.8 Other CUSUM charts

(a) Control of process variability

If our data are normally distributed and we wish to control process variability, then CUSUM charts can be used for this. If observations occur singly, a CUSUM chart can be operated on absolute successive differences,

$$d_i = | x_{i+1} - x_i |.$$

If the data are grouped, a CUSUM chart can be operated on ranges, sample variances, or on log (sample variance).

The ARL properties of these various procedures have not been obtained, nor have any comparative studies been made.

(b) Discrete data

CUSUM charts for binomial and Poisson data have been considered by Ewan and Kemp (1960), who also provided information on ARL properties of the procedures. A CUSUM scheme for binomial data had been considered earlier by Page (1954).

The intermediate case where data are gauged and assigned a score is considered by Page (1962).

(c) *Other distributions*

Johnson (1966) has considered the use of CUSUM charts to control the mean of a Weibull distribution.

4.9 Conclusions

One of the vital questions about which we have said very little is a comparison between, say, the Shewhart $\bar{x}$-chart and a CUSUM chart. It is clear from Figure 4.8 and surrounding discussion that there are occasions when each will be preferred, on a basis of the ARL alone. It depends largely whether small or large changes in the mean are of primary interest.

There are other considerations which may influence the choice of chart to be used in any practical case. An $\bar{x}$-chart may be thought simpler than a CUSUM chart. Roberts (1966), after a comparative study, concludes: 'We suggest that most applications are and will continue to be of such a nature as to favour the use of the standard $\bar{x}$ and range control charts. . . .' On the other hand, there are many large industrial establishments which have changed over completely to CUSUM charts. Ewan (1963) gives an excellent discussion of various control charts. He suggests that standard control charts should be used when extreme simplicity in use is required, or when tests are very inexpensive so that improvements in efficiency are of small importance. CUSUM charts are more appropriate when tests are moderately or very expensive and extreme simplicity in use is not so vital, and when it is required to detect quickly sudden but sustained changes in the mean. CUSUM charts also show up to considerable advantage when it is of value to have an estimate of the point at which a change in the mean occurred.

The paper by Ewan (1963) is very simple, and should be read by all interested in this topic.

5. Continuous sampling plans

5.1 Basic description and aims

In section 1.3(*a*) we drew a distinction between batch inspection and continuous production inspection, and we explained that the latter deals with the inspection of either truly continuous material such as nylon thread, or else of conveyorized production of separate items such as chocolate bars. The material of Chapters 3 and 4 relates to this situation. However, a special set of inspection plans, known as continuous sampling plans, has been introduced for use in continuous production inspection.

The earliest CSP, introduced by Dodge in 1943, has already been described in Example 1.4, and this plan is referred to as CSP-1. In CSP-1 there are two levels of inspection, 100% inspection and an inspection rate of $1/n$, and there is a simple rule to determine when to change between these levels. Variations on this basic plan are either to use a more complex rule for changing inspection levels, or else to introduce more levels.

One possible approach to continuous production inspection is to group the product artificially in batches. It is frequently necessary to group the material for transit purposes; these groups could be used as batches for inspection. However, any artificial batching may have unfortunate results. Firstly, the operation of artificially batched sampling plans can lead to the possibility of rejecting items not yet produced. Secondly, when inspection involves disassembly, or is time-consuming, many practical difficulties arise, such as storage problems. Nevertheless, artificial batching of continuous output is used as a method of reducing the problem of designing inspection plans to that described in Chapter 2. In the present chapter we discuss sampling plans suitable when artificial batching is not appropriate.

A producer operating a continuous sampling plan such as CSP-1 may have any or all of three different aims in view:

(*i*) *Product screening*. The aim in this case has been emphasized throughout Chapter 2. The product is to be sorted, usually into two grades, an acceptable grade and one which needs to be rejected or rectified.

(*ii*) *Process trouble shooting*. This was discussed in section 3.1. Typically, the assumption is that product quality is occasionally disturbed by 'assignable causes of variation', which can be traced and eliminated.

(*iii*) *Adaptive control*. Here the inspection results are to be used to indicate the precise amount of any adjustment needed to the process in order to keep quality up to standard.

The original work by Dodge (1943), and much work since, such as Dodge and Torrey (1951), Lieberman and Solomon (1955), has emphasized product screening although process trouble shooting is also in view. The term adaptive control was used by Box and Jenkins (1962, 1963), but some earlier work by Girshick and Rubin (1952), Bishop (1957, 1960) and a large literature on control theory is relevant. Savage (1959) designed a plan specifically for trouble shooting. General reviews of the literature are given by Bowker (1956), Chiu and Wetherill (1973), Duncan (1974, chapter 17), Lieberman (1965), and Phillips (1969).

Exercises 5.1

1. Discuss how CUSUM charts might be used for the continuous sampling problems mentioned in this section.

5.2 CSP-1 and the AOQL criterion

It is convenient here to restate the CSP-1 sampling plan.

CSP-1. Inspect every item until i successive items are found free of defects, and then inspect at a rate of one in every nth item. When a defective item is found, revert to 100% inspection, and continue until i successive items are found free of defects. □□□

Dodge required the sampling at a rate 1 in n to be carried out by stratified random selection so as to ensure an unbiased sample. In practice inspectors are likely to select approximately every nth item, but it is wise to vary this interval a little.

The way that the CSP-1 and similar plans operate is to vary the inspection rate as quality varies. Clearly, a theoretical model is required to give a guide on how the inspection rate varies with p, for various choices of n and i.

In most theoretical treatments of CSP-1 the following three assumptions are made.

Assumption (1). All defectives found during inspection are rectified or replaced by good items.

Assumption (2). Inspection is perfect, i.e. mistakes in identifying defectives are never made.

Assumption (3). Theoretical calculations are made on the assumption that the process is producing defectives with probability p, and that the probability that any item is defective is independent of the quality of other items.

Assumption (1) is often realistic, but if it is not, account of this can be taken in the theory. Assumption (2) is unrealistic and we shall have to discuss this later. Assumption (3) is effectively that the process is in a steady state and provided that we realize the implications, it is realistic enough to proceed with some simple theory.

In the next section we show that on these three assumptions, the average fraction of production inspected is

$$F(p) = 1/\{1 + (n-1)q^i\} \tag{5.1}$$

where $q = 1 - p$. On Assumption (1), the average outgoing proportion defective is therefore

$$\text{Outgoing proportion defective} = p\left\{1 - \frac{1}{\{1 + (n-1)q^i\}}\right\}$$

$$= \frac{p(n-1)q^i}{\{1 + (n-1)q^i\}}. \tag{5.2}$$

It should be stressed that this formula assumes a constant p; if p has, say, a cyclic variation, quite a different result will hold.

Now (5.2) has approximately the shape shown in Figure 5.1. For low p, the outgoing proportion defective is low. For high p, the average fraction of production inspected is high, and again the outgoing proportion defective is low. For intermediate values of p, there is a maximum value to the average outgoing proportion defective for a given n and i, and this is defined as the average outgoing quality limit, AOQL.

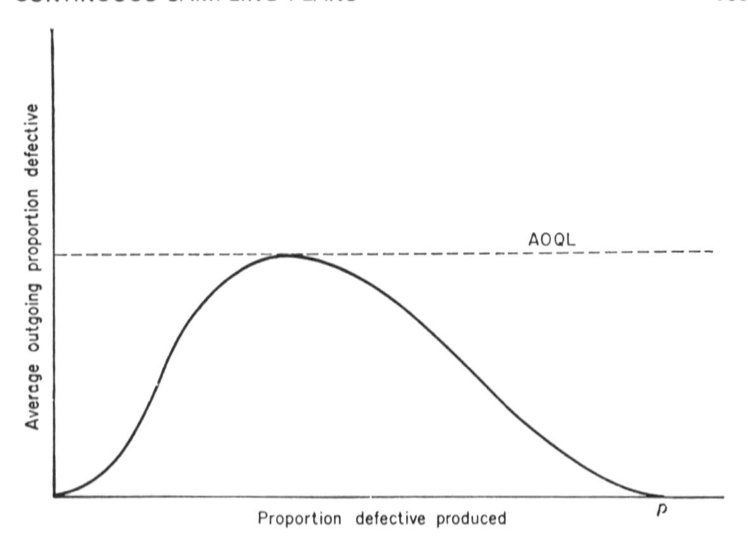

Figure 5.1. *Operation of CSP-1.*

The AOQL is the maximum of (5.2), and by differentiation we find that this is at a value $p = p_1$, where

$$(i + 1)p_1 - 1 = (n - 1)(1 - p_1)^{i+1} \tag{5.3}$$

and by inserting this we find that the AOQL is

$$\text{AOQL} = (n - 1)(1 - p_1)^{i+1}/i, \tag{5.4}$$

which can be regarded as a function of n and i.

Figure 5.2 shows approximately how the AOQL is related to n and i. Dodge suggested that a producer be asked to specify an AOQL, so that this sets a relationship between n and i. The final choice of n and i was to be made on practical considerations such as the work load on inspectors, and it may be best to have an i no greater than a small multiple of the number of units on the production line at any time.

This method of designing a CSP-1 has certainly been used a great deal since Dodge suggested it. However, let us reflect on how artificial is the concept of the AOQL:

(*i*) The AOQL is an upper limit to the proportion defective only in a long-run average sense. In the short run, a sudden deterioration of quality could lead to a large number of defectives being passed before a defective was found on inspection. This should clearly be borne in mind when choosing n; see section 5.4.

(*ii*) We have made Assumption (3), that the process is in control.

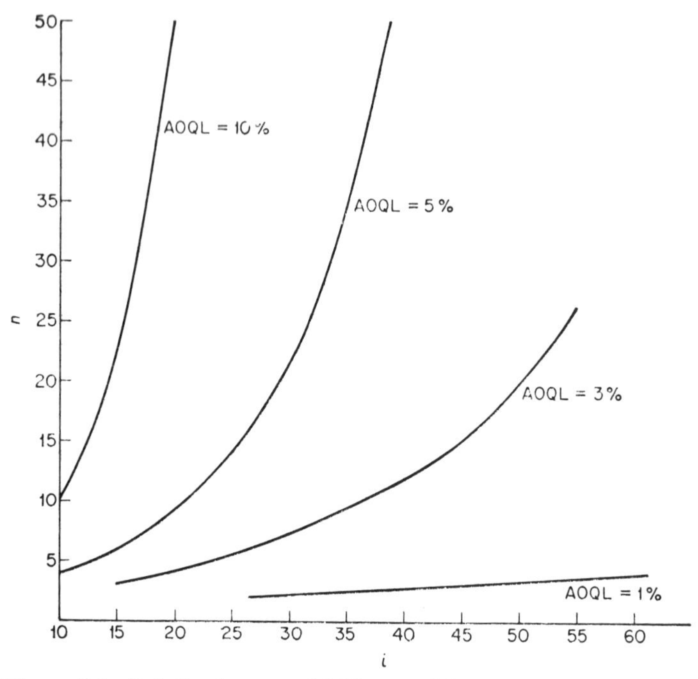

Figure 5.2. *Relation between AOQL, n and i.*

If the process has varying quality, with changes exactly in phase with changes in the inspection level, the AOQL no longer applies.

(*iii*) The quality, p, of the uninspected production process at which the AOQL is obtained may be known to occur only very rarely.

(*iv*) We have made Assumption (2). If defective items are only recognized with a probability of, say 0·90 or 0·95, Figure 5.1 does not apply, and instead we have Figure 5.3. This situation is therefore likely to make nonsense of the whole AOQL concept. Hill (1962) has stressed that the AOQL concept is particularly sensitive to the assumption that inspection is perfect.

Notwithstanding these criticisms, it should be emphasized that the CSP-1 has been successfully designed and used in the way Dodge suggested, although there is clearly a need for other design criteria.

Exercises 5.2

1. Check the derivation of (5.3) and (5.4) from (5.2).

2*. Examine how this theory is altered when inspection is imperfect, and find the conditions under which there is a true maximum to the

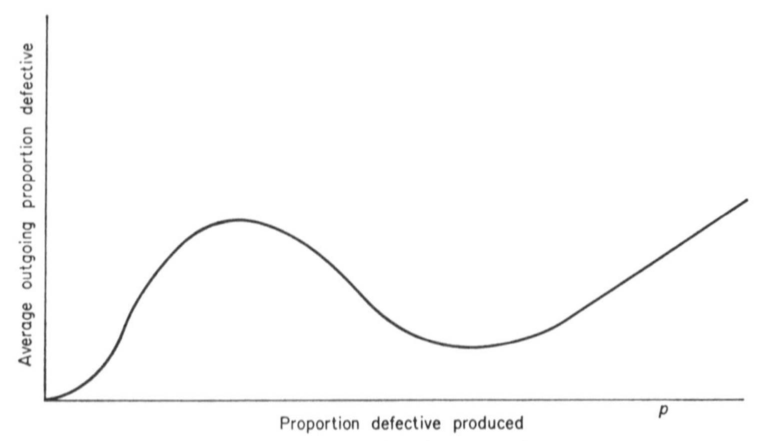

Figure 5.3. *The CSP-1 when inspection is not perfect.*

average outgoing quality.

5.3 Theory of CSP-1

In this section we derive the theory of CSP-1 on the three assumptions stated in the previous section.

The first step is to break up the run of inspected items at every defective. Dodge calls these short sequences 'terminal defect sequences', and the following are examples; O stands for a good item and X for a defective:

Sequence	*Probability*	*Length*
X	p	1
OX	pq	2
OOX	pq^2	3
. . .	. . .	. . .
OO . . . OX	pq^i	$(i + 1)$
. . .	. . .	. . .

Once a defective has been observed, 100% inspection is started and continued until a sequence of i good items is observed. Before this occurs, a series of terminal defect sequences may occur of length less than or equal to i. The probability of a terminal defect sequence of length less than or equal to i is

$$\sum_{r=0}^{i-1} pq^r = 1 - q^i = P, \quad \text{say.} \tag{5.5}$$

The number of terminal defect sequences in a run of 100% inspection has a geometric distribution $(1 - P)P^r$, $r = 0, 1, 2, \ldots$; the average number of such sequences is therefore

$$E(l) = \sum r(1 - P)P^r = P/(1 - P) = (1 - q^i)/q^i. \tag{5.6}$$

Now the average length of a terminal defect sequence of length less than or equal to i is

$$T = \frac{1}{(1 - q^i)} \sum_{r=0}^{i-1} (r + 1)pq^r = \frac{\{1 - q^i(1 + pi)\}}{p(1 - q^i)}. \tag{5.7}$$

The average length of a run of 100% inspection is therefore

$$TE(l) + i = (1 - q^i)/pq^i. \tag{5.8}$$

The number of periodic samples taken in between runs of 100% inspection has the geometric distribution pq^{r-1}, $r = 1, 2, \ldots$. The average number of items passed in such an interval is therefore

$$n \sum rpq^{r-1} = n/p. \tag{5.9}$$

The average number of items produced between the start of successive runs of 100% inspection is therefore

$$(1 - q^i)/pq^i + n/p$$

while the amount inspected in such an interval is

$$(1 - q^i)/pq^i + 1/p.$$

The average fraction inspected is therefore

$$F = \frac{(1 - q^i)/pq^i + 1/p}{(1 - q^i)/pq^i + n/p} = \frac{1}{1 + (n - 1)q^i}. \tag{5.10}$$

This is the formula quoted in (5.1), from which (5.2), (5.3) and (5.4) follow.

When interpreting this result, reference should be made to the criticisms listed at the end of the previous section.

5.4 The AEDL criterion

In section 5.2 we remarked that if there was a sudden deterioration of quality, a number of defective items could be passed by the CSP-1 before 100% inspection was instituted. Hillier (1964) proposed another measure, the AEDL or Average Extra Defectives Limit, the purpose of which is to put a limit on the average number of defectives passed upon such a deterioration of quality. The AEDL criterion can be used along with the AOQL to select a particular CSP-1.

Suppose a process is producing defectives with probability p_0, and suddenly it changes to producing defectives with probability $p_1 > p_0$. Let D be the number of uninspected defectives among the next L items after this deterioration of quality. Then for an AOQL of θ, the average extra number of defectives passed above the limit prescribed by the AOQL is

$$\{E(D) - \theta L\}$$

and this will be a function of p_0, p_1, and L. The AEDL, written D_L, is defined as

$$D_L = \max_{p_0, p_1, L} \{E(D) - \theta L\}. \tag{5.11}$$

For the CSP-1, Hillier shows that (5.11) achieves its maximum for $p_0 = 0$, $p_1 = 1$, and $L = L^*$, where

$$L^* = \log \left\{(1 - n) \log \left(\frac{n-1}{n}\right) \Big/ \theta \right\} \Big/ \log \left(\frac{n-1}{n}\right). \tag{5.12}$$

Hillier shows that the AEDL for CSP-1 is then

$$\begin{aligned} D_L &= (n-1)\left\{1 - \left(\frac{n-1}{n}\right)^{L^*}\right\} - \theta L^* \quad &\text{if } L^* > 0 \\ &= 0 &\text{if } L^* \leq 0. \end{aligned} \tag{5.13}$$

In a particular case there may be reason to use a value of L other than (5.12), for example, if items are packaged in batches of a given size. This draws attention to the fact that the AEDL is a number of defectives calculated over a somewhat arbitrary length of production.

The other criticisms of the AOQL criterion made at the end of section 5.2 will be found to apply also to the AEDL. In particular the values $p_0 = 0$ and $p_1 = 1$ at which the AEDL is calculated are both rather unlikely values for the proportion defective. However, the use of the AEDL together with the AOQL would seem to be a better method of choosing a particular CSP-1 than the use of the AOQL alone, and Hillier gives a simple example. The AEDL provides a method of choosing n, by using (5.13).

Hillier suggests that this method of choosing a CSP-1 can be improved further if account is taken of the probability distribution of D for given values of L. It would then be possible to make the probability that D is less than a given number to be greater than a specified value. See Hillier (1964) for details of this method. Unfortunately there is very little published information on the probability distribution of D; see Hillier (1961, 1964).

5.5 Decision-theory approach to CSP-1

Anscombe (1958) gives a critique of the AOQL approach to choosing a CSP-1, and discusses an approach based on costs. He points out that the AOQL concept is very artificial, and would not usually correspond to what a user required of a continuous sampling plan. The problem is basically an economic one of balancing inspection costs against the costs of passing defective items. The usual objection to an economic approach is that the cost data may be difficult to obtain. However, Anscombe says: 'What is important is that we realize what the problem really is, and solve that problem as well as we can, instead of inventing a substitute problem that can be solved exactly but is irrelevant.' If the cost of passing defectives is known only roughly, then an approximate solution to the problem will be satisfactory, provided we are solving the real problem.

Admittedly, there are other aims in inspection besides the strict economic aim of limiting the amount of bad material passed, but this aim is likely to be the over-riding one. The approach adopted by Anscombe requires very little economic information, but this small amount is vital.

We shall again make the three assumptions listed in section 5.2. Let the cost of inspection be k cost units, where the unit of costs is the excess cost of passing a defective item above the cost of rectifying it or replacing it during inspection. The cost of 100% inspection is therefore k per item produced, and the cost of passing production without inspection is p per item produced. By this model therefore, it would be best to carry out 100% inspection if $k < p$, and best not to inspect at all if $k > p$. In practice the proportion defective p varies, and a sampling plan is operated.

If we operate a CSP-1, the cost of this plan per item produced is

$$\begin{aligned} C &= (\text{cost of inspection}) + (\text{cost of passing defective items}) \\ &= k \times (\text{fraction inspected}) + p \times (\text{fraction not inspected}) \end{aligned}$$

or $C = Fk + (1 - F)p$ (5.14)

where F is given by (5.1). As indicated above, the best possible action if we knew p is

$$\text{for } p < k, \quad C = p \quad (F = 0 \text{ in } (5.14))$$

and

$$\text{for } p > k, \quad C = k \quad (F = 1 \text{ in } (5.14)).$$

The excess cost ΔC over the best possible action is therefore

$$\Delta C = \begin{cases} (k - p)F & p < k \\ (p - k)(1 - F) & p > k. \end{cases} \qquad (5.15)$$

Anscombe now simplifies the problem by inserting an arbitrary rule which appears to be near optimum. Since the best possible action changes from no inspection to 100% inspection at $p = k$, it is reasonable to choose $F = \frac{1}{2}$ when $p = k$. By inserting this rule into (5.1) we obtain

$$(n - 1)(1 - k)^i = 1, \tag{5.16}$$

which can be used to calculate i for a given n and k. If k is given, there remains only one parameter of the CSP-1, namely n, which we wish to optimize.

The next step is to find the average value of ΔC over the process curve for p, assuming that p varies slowly enough for (5.2) to remain valid. Anscombe introduced a further approximation here, by using a uniform distribution for p in the range $(0, 2k)$. (If the variation of p does not span the point $p = k$, the optimum will be either no inspection or 100% inspection. Furthermore, it turns out that for a wide range of distributions the value of $E(\Delta C)$ obtained is not very different from that obtained under the uniform distribution.) By numerical integration, Anscombe now checks that the empirical formula

$$E(\Delta C) = 0{\cdot}3k/\sqrt{n} \tag{5.17}$$

holds very well.

In order to be able to determine an optimum for n we must introduce one further factor in the costs. Equation (5.17) will give an approximation to the long-run costs of a CSP-1 at a stable value of p. When p changes, a further cost arises, called a transition cost. This is the cost of the extra defectives passed after a sudden deterioration of quality, and before the CSP-1 changes to 100% inspection. If p changes suddenly from a very small value to a very large value, at a random point in an inspection interval, then on average slightly less than $n/2$ defectives will be passed. Anscombe showed that $n/2$ is a good approximation to the average transition cost under more general conditions.

If sudden deteriorations of quality occur on average once in every M items produced, the average transition costs are $n/2M$ per item produced.

The total cost of operating the CSP-1 is therefore approximately

$$\frac{0{\cdot}3k}{\sqrt{n}} + \frac{n}{2M} \tag{5.18}$$

and by differentiating we find that the optimum choice of n is

$$n = (0{\cdot}3kM)^{3/2}. \tag{5.19}$$

In obtaining this result we have used the rule $F = \frac{1}{2}$ at $p = k$, the uniform distribution as an approximation to the process curve, the empirical approximation (5.17), and the approximations to the transition costs. Further investigation shows that none of these approximations have much effect on the solution. The important quantities are M, the average interval between sudden deteriorations of quality, and

$$k = \frac{\text{cost of inspecting an item}}{\text{excess cost of passing a defective}}.$$

It is interesting that in the methods suggested earlier in this chapter for choosing a CSP-1, neither k nor M were mentioned, and these are the quantities upon which an optimum solution strongly depends.

Exercises 5.5

1. Examine numerically the relationship (5.16) for $k = 0{\cdot}05$.
2. Show that an approximation to (5.16) is

$$ki \simeq \log_e n.$$

3*. Examine the effect on this theory of inspection being imperfect.

5.6 Modifications to CSP-1

Over the years various modifications have been suggested to CSP-1. Dodge and Torrey (1951) suggested the following two plans:

CSP-2. Proceed as in CSP-1 except that, once partial inspection is instituted, 100% inspection is only introduced when two defectives occur spaced less than k items apart. □□□

This plan is less likely to revert to 100% inspection because of isolated defectives than is the CSP-1, and the number of abrupt changes of inspection level will also be reduced. However, there is a higher risk of accepting short runs of poor quality, and so CSP-3 is suggested.

CSP-3. Proceed as in CSP-2 except that when a defective is found, the next four items are inspected. □□□

The theory of these two plans follows a similar pattern to the theory given in section 5.3 for CSP-1, although in each case it is more complicated.

Another line of development attempts to devise plans which

guarantee an AOQL without assuming statistical control of the process. The starting-point of these investigations is a paper by Lieberman (1953), who examined the AOQL of the CSP-1 without the assumption of control. It is not difficult to see that this is attained by a process which produces good items throughout periods of 100% inspection, and defectives throughout periods of partial inspection. Periods of 100% inspection are therefore exactly i items long, and the average number of items produced between the start of such periods is $(n + i)$. One defective item will be inspected, and consequently replaced by a good item. The average fraction defective remaining after inspection is therefore $(n - 1)/(n + i)$, which can be considerably greater than (5.2). For a formal proof of this formula, see Derman *et al.* (1959). When interpreting this result, however, it is important to take note of the pathological nature of the production process model which produces it.

Derman *et al.* (1959) present two variants of CSP-1 which have improved properties when control is not assumed.

CSP-4. Proceed as in CSP-1 except that partial inspection is carried out by separating production into segments of size n, and taking one item at random from each segment. When a defective is found, the remaining $(n - 1)$ items in the segment are eliminated from the production process, and 100% inspected started with the first item of the following segment. □□□

The idea of CSP-4 is that there is a reluctance to pass a segment of production in which a defective is found. Items eliminated from the production process might be sorted and the good items used as a stock for replacing defectives found in inspection. A more realistic plan would be to allow the good items from this 'eliminated' segment to be passed, and so we have CSP-5.

CSP-5. Proceed as in CSP-4 except that all items in a segment in which a defective is found are sorted. □□□

The modifications given in CSP-4 and CSP-5 result in a more complicated set-up when control is not assumed. The production process model giving the AOQL is no longer the trivial one described earlier for CSP-1. The theory is not simple, and we refer readers to the source paper. In practice, Derman *et al.* (1959) suggest that CSP-4 and CSP-5 plans should be chosen using the CSP-1 formula derived under the assumption of control.

Another important type of plan is the multilevel plan, discussed by Lieberman and Solomon (1954), and we shall designate this MLP-1.

MLP-1. Proceed as in CSP-1 except as follows. If in partial inspection i successive items are found free of defects, reduce the inspection rate from $1/n$ to $1/n^2$. In this way, several inspection levels can be used. When a defective is found, revert to 100% inspection.

□□□

Usually MLP-1 will be used with between two and six levels. Lieberman and Solomon (1954) obtained the AOQL for two levels and for an infinite number of levels, and gave a method of interpolation for other levels. Clearly, a whole range of different types of multilevel plan is possible, but no systematic study of the possibilities seems to have been undertaken.

In nearly all of the work an AOQL approach is adopted, and the AEDL criterion has only been applied to CSP-1. Anscombe's decision-theory approach, described in section 5.5, has not been extended to cover other plans. That is, with very few exceptions, Dodge's original formulation of the continuous inspection problem has not been questioned.

Read and Beattie (1961) give a plan of the same general type as CSP-1, but modified to fit their practical conditions. The inspection rate on line is held constant, and the product is artificially batched. Depending on the results of inspection, some batches are set aside for 100% inspection later. This plan forms a link between the Dodge type continuous inspection plans, and batch inspection plans discussed earlier.

A collection of continuous sampling plans, indexed for use as a United States Army military standard, is available as MIL-STD-1235 (ORD). This standard is currently being revised, and for a description and discussion of the revision principles see Banzhaf and Bruger (1970), Duncan (1974), and Grant and Leavenworth (1972).

Exercises 5.6

1. Make the three assumptions listed in section 5.2, and find the formulae equivalent to (5.1) and (5.2) for CSP-2 and CSP-3, when $k = i$. See Bowker (1956).

5.7 Process trouble shooting

So far we have been concentrating mostly on the product screening aspect of continuous inspection. Girshick and Rubin (1952), in an important paper, gave a Bayes approach to process trouble shooting, and we briefly describe their theory below.

The production process is assumed to be either in a good state (state 1), or a bad state (state 2). After every item produced there is a probability g that the process will move from states 1 to 2, but once in state 2, the process remains in that state until it is brought to repair. Girshick and Rubin derive an optimum rule for deciding when to put the process in repair. If the process is put into repair when it is in state 1, it is said to be in state 3, and if it is put into repair from state 2, it is said to be in state 4. When the process is put into states 3 or 4, it remains there for n_j time units, $j = 3, 4$, where one time unit is the time for one item to be produced. Two cases are considered:

(*i*) 100% inspection is operated and the problem is merely to find the optimum rule for deciding when to put the process in repair.

(*ii*) Sampling inspection can be used, so that the optimum rule must also specify when items are to be inspected.

These two cases are discussed separately below.

The quality of each item produced is represented by a variable x, and the probability density function of x is taken to be $f_j(x)$, $j = 1, 2$, for states 1 or 2 respectively. The value of an item of quality x is $V(x)$, and the cost per unit time of the repair states is c_j, $j = 3, 4$. The model is now precisely defined, and we have to find the decision rules which maximize income per unit time. This model is sufficiently general and realistic to be used as a means of comparing various continuous inspection procedures, but no such comparisons have yet been made.

When the production process is in use, the vital question is to decide whether it is in state 1 or state 2. Clearly, the optimum decision rule will depend on the posterior probability that the next item will be produced in state 1. For case (*i*) above and when the kth item has just been inspected this probability is clearly

$$q_k = \frac{(1 - g)q_{k-1}f_1(x_k)}{q_{k-1}f_1(x_k) + (1 - q_{k-1})f_2(x_k)} \tag{5.20}$$

where $q_0 = 1 - g$. (The denominator is the probability that x_k is observed, and the numerator is the probability that x_k is produced in state 1, and that the process remains in state 1 for the $(k + 1)$th item.)

Girshick and Rubin showed that the optimum rule is to put the process in repair whenever $q_k \leq q^*$. This is equivalent to putting the process in repair whenever $Z_k \geq a^*$, where

$$Z_k = y_k(1 + Z_{k-1}), \qquad Z_0 = 0, \tag{5.21}$$

and

$$y_k = f_2(x_k)/\{(1 - g)f_1(x_k)\}. \tag{5.22}$$

The parameter a^* has to be chosen to maximize income per unit time, and this involves solving an integral equation.

When sampling inspection can be used, the argument and result are very similar. The optimum rule is again defined in terms of Z_k, where y_k is given by (5.21) if the kth item is inspected and

$$y_k = (1 - g)^{-1} \tag{5.23}$$

if the kth item is not inspected. Girshick and Rubin show that the optimum rule is to inspect items whenever

$$b^* \leq Z_k < a^*,$$

to put the process in repair when $Z_k \geq a^*$, and to pass production without inspection whenever $Z_k < b^*$. Again the constants b^* and a^* have to be chosen to maximize income per unit time, and this involves solving integral equations.

In both cases the integral equations are very difficult to solve, and detailed calculations do not appear to have been carried out.

Exercises 5.7

1. Find q_k in terms of Z_k and g.

2*. In the above theory, the quality x of each item is assumed to be observed exactly. What happens if the quality of each item is observed with error?

5.8 Adaptive control

There is now a very large literature on control theory, and this volume would be incomplete without a brief introduction to it. Those interested in pursuing the topic further should read the general accounts by Barnard (1959), Lieberman (1965), and White (1965), and the references contained in these papers. The following account is largely based on Box and Jenkins (1962).

Suppose a process is sampled at equal time intervals, and that provided no adjustments are made to the process the observation at the jth sample point is

$$z_j = \theta_j + u_j,$$

where u_j are the errors which are normally and independently distributed with a variance σ_u^2, and θ_j follows some stochastic process.

Adjustments can be made to the process at each sample point, and the aim of these adjustments is to keep θ_j at a target value, which we may without loss of generality take to be zero. If the total adjustment applied at the jth sample point is X_j, the observation made is the apparent deviation from the target value, which is

$$e_j = z_j - X_j = \theta_j - X_j + u_j = \epsilon_j + u_j,$$

where ϵ_j is the actual deviation from the target value.

Suppose adjustments have been made on some basis or other, and that we have data $X_1, X_2, X_3, \ldots, X_j$, and $e_1, e_2, e_3, \ldots, e_j$, then our problem is to determine the increment x_{j+1} to apply to the adjustment at the $(j+1)$th sample point, so that the total adjustment is then

$$X_{j+1} = X_j + x_{j+1}.$$

We are assuming, of course, that adjustments can be applied at every sample point without extra cost.

Let the loss caused by an actual deviation from target of ϵ_j be proportional to ϵ_j^2, then we must determine x_{j+1} so that $E(\theta_{j+1} - X_{j+1})^2$ is minimized. If we take x_{j+1} to be a linear function of $e_j, e_{j-1}, \ldots$, this means that we must determine x_{j+1},

$$x_{j+1} = \hat{\theta}_{j+1} - \hat{\theta}_j = \sum_{r=0}^{\infty} w_r e_{j-r} \tag{5.24}$$

where the w_r's are chosen so that $\hat{\theta}_{j+1}$ is the minimum mean square error estimate of θ_{j+1}. In fact the central problem as stated here is seen to be equivalent to the problem of predicting the coming value of θ_{j+1}. The problem can therefore be restated as the problem of determining weights μ_r so that

$$\hat{\theta}_{j+1} = \sum_{r=0}^{\infty} \mu_r z_{j-r} \tag{5.25}$$

is the minimum mean square error predictor of θ_{j+1}. (Again, a linear function is assumed for simplicity.) This implies, of course, a relationship between the w_r's and the μ_r's.

So far we have said nothing about the stochastic process to be assumed for θ_j, and it would be unrealistic to assume that it was stationary. Suppose that θ_j can be separated into two components,

$$\theta_j = m_j + \phi_j,$$

where m_j is a sequence of known means, and where ϕ_j is a first-order autoregressive process,

$$\phi_{j+1} = \rho\phi_j + \eta_j$$

where the η_j are independently and normally distributed with a variance σ_η^2. In a practical case the m_j would not be known, but we first obtain the optimum weights assuming them to be known.

A further simplification is introduced by assuming the weights μ_r to be zero for $r \geq h$, for some specified h. With these assumptions the covariance matrix of $z_j' = (z_j, z_{j-1}, \ldots, z_{j-h+1})$ is

$$\mathbf{T} = \begin{pmatrix} \rho\sigma_\theta^2 + \sigma_u^2 & \rho^2\sigma_\theta^2 & \rho^3\sigma_\theta^2 & \cdots \\ \rho^2\sigma_\theta^2 & \rho\sigma_\theta^2 + \sigma_u^2 & \rho^2\sigma_\theta^2 & \cdots \\ \vdots & \vdots & & \end{pmatrix}$$

where $\sigma_\theta^2 = \sigma_\eta^2/(1 - \rho^2)$.

Box and Jenkins (1962) now show that the weights μ_j which give the minimum mean square error predictor are

$$\boldsymbol{\mu} = T^{-1}\boldsymbol{\rho} \tag{5.26}$$

where $\boldsymbol{\mu}' = (\mu_0, \mu_1, \ldots, \mu_{h-1})$ and $\boldsymbol{\rho}' = (\rho, \rho^2, \ldots, \rho^h)$, and where we use the estimate

$$\theta_{j+1} = m_{j+1} + \sum_{r=0}^{h} \mu_r(z_{j-r} - m_{j-r}). \tag{5.27}$$

Now if the m_j are not known, we shall have to use the estimate (5.25), and there will be a bias. However, if the m_j follows a polynomial of degree k, constraints can be imposed on the weights μ_r so that the bias is zero. The optimum weights can now be found subject to these constraints, but the result is rather complicated to state, and we refer the reader to Box and Jenkins (1962). The authors evaluate the optimum constrained predictors for some simple cases, and show that they are such that a good approximation to the optimum change x_{j+1} is

$$x_{j+1} = \gamma_{-1}\,\Delta e_j + \gamma_0 e_j + \gamma_1 \sum_{r=0}^{\infty} e_{j-r}, \tag{5.28}$$

or a simple generalization of it. Box and Jenkins then examine the stochastic process for which an adjustment of the type (5.28) would be optimum, and they consider methods of estimating the parameters of this process from data. All this theory therefore leads to the following empirical approach; a process model is fitted to past data, so determining a set of parameters $\gamma_{-1}, \gamma_0, \gamma_1, \ldots$, and then an adjustment of the type (5.28) is used, inserting the fitted parameters.

The discussion in Box and Jenkins (1962) is more general than the discussion given above, but the authors state that some of the more general results are unlikely to be used because of their complexity. In a subsequent paper, Box and Jenkins (1963) again consider the above problem, but with the introduction of a cost for being off target and a cost for making a change; the optimum plan then involves making adjustments to the process less frequently.

Further developments would be of interest. For example, it may be desirable to vary the inspection rate depending upon the results. Another point which does not seem to be adequately cleared up is the relationship of the methods suggested in this section to adaptive control by CUSUM methods, and some remarks by Barnard in the discussion of Box and Jenkins (1962) relate to this. Barnard suggests that CUSUM methods may be preferred because of simplicity in cases where computers are not available to do the calculations, but that in certain circumstances, CUSUM methods may be slightly better anyway.

5.9 Use of CUSUM techniques

A general question is opened up by the closing remarks of the last section, relating to the possibility of basing continuous sampling plans on CUSUM techniques. One such plan is given by Beattie (1968) in an important paper dealing with patrol inspection, when an inspector is asked to cover a large area of a factory taking small samples.

One plan proposed by Beattie (1962, 1968) is as follows. The inspector makes periodic inspections and on each occasion he selects n items, finding d_i defectives, $i = 1, 2, \ldots$. A CUSUM is now plotted for $\sum (d_i - k)$, where k is some reference value, as shown in Figure 5.4. The stream of product is accepted while plotting is on the lower chart. When the plot on the lower chart reaches the decision interval, the product is rejected, and plotting is started on the upper chart. Plotting on the upper chart continues until the decision interval is reached, when the stream of product is again accepted, and plotting on the lower chart restarted.

The rejected product is separated into lots, and a single sample plan applied to each lot. The plan therefore operates in rather a similar way to the CSP-1. In general only periodic samples of size n are taken, but periods of acceptance sampling of lots are required, when quality deteriorates.

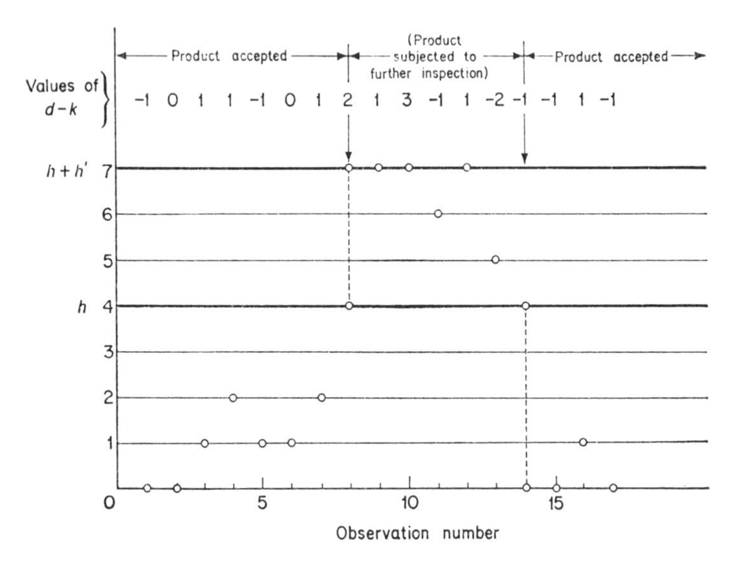

Figure 5.4. *A two-stage semi-continuous plan.* (*A combination of CUSUM charts.*)

Clearly, when acceptance sampling is being operated, a double or sequential sampling plan can be used instead of a single sampling plan.

For theoretical purposes let us suppose that production is artificially separated into lots, and that m such lots pass in between inspection periods by the patrol inspector. In calculating ARL's, we shall use this lot size as a unit.

The CUSUM chart just described is slightly different from the two-sided CUSUM chart described in section 4.2, in that only one chart is used at a time. Let z be the score on the lower chart and $L(z, p)$ be the ARL for a starting score of z, where p is the proportion defective. Then by following the discrete analogue of (4.8) we have

$$L(z, p) = 1 + L(0, p) \sum_{x=0}^{k-z} f(x) + \sum_{x=1}^{h-1} L(x) f\,(y+k-z) \qquad (5.29)$$

where $$f(x) = {}^{n}C_{x}\, p^{x}(1-p)^{n-x}$$

which is the probability that x defectives are found in the first sample of size n. Equation (5.29) can be solved to obtain $L(0, p)$. Similarly we can obtain the ARL $L'(0, p)$ of the upper chart. The probability P, that lot inspection is not used, is then seen to be

$$P_1(p) = L(0, p)/\{L(0, p) + L'(0, p)\}. \tag{5.30}$$

If the lot inspection plan leads to acceptance with a probability $P_2(p)$, the total probability of acceptance is

$$P_a(p) = P_1(p) + \{1 - P_1(p)\}P_2(p). \tag{5.31}$$

If the sample size for the lot inspection plan is n', the average sample number per lot inspected is

$$\text{ASN} = n/m + n'(1 - P_1). \tag{5.32}$$

Expressions (5.31) and (5.32) are functions of n, n', h, h', k, p, and the acceptance number for the lot inspection plan. In choosing a particular plan Beattie suggests using OC-curve considerations, together with consideration of the ASN at the expected quality level. However, there are clearly other schemes for choosing a particular plan, and this aspect does not appear to have been thoroughly investigated.

For further work on this type of use of CUSUM charts see Beattie (1968), Prairie and Zimmer (1970), and Rai (1971); the last two of these references relate to inspection by variables.

Papers and books for further reading

General reference

Duncan (1974). Includes numerical examples, exercises, and sampling tables.

Chapter 2 Acceptance sampling

Practical background: Hill (1962). Papers by Tippett and Hamaker in *Applied Statistics*, 8, 1956. Chiu and Wetherill (1975).
Serial sampling schemes: Cox (1960).
Double sampling: Hamaker and Van Strik (1955). Wetherill and Campling (1966).
Multiple sampling plans: Wilson and Burgess (1971).
Theoretical papers: Barnard (1954). Guenther (1971). Hald (1960). Horsnell (1957). Page (1954).
Review paper: Wetherill and Chiu (1975).
Inspection by variables: Fertig and Mann (1974). Lieberman and Resnikoff (1955). Owen (1964, 1967, 1968, 1969).

Chapter 3 Control charts

Practical background: Grant and Leavenworth (1972).
Specifications for variables: Hill (1956).
Theoretical: Duncan (1956). Page (1955). Schmidt and Taylor (1973). Also see King (1952).

Chapter 4 Cumulative sum charts

Simple exposition: Kemp (1962).
Practical background: Ewan (1963). Traux (1961). Woodward and Goldsmith (1964).
Deviations from assumptions: Bissell (1969).
Comparison of charts: Ewan (1963). Roberts (1966).
Derivation of properties: Brook and Evans (1972). Kemp (1971). De Bruyn (1968).
Theoretical papers: Barnard (1959). Taylor (1968).

Chapter 5 Continuous sampling plans

See the references at the end of section 5.1, the beginning of section 5.8, and the end of section 5.9.
General description: Dodge (1970). Grant and Leavenworth (1972).
Multilevel continuous plans: Derman *et al.* (1957). Lieberman and Solomon (1955). Sackrowitz (1972).
Economic viewpoint: Chiu and Wetherill (1973). Kao (1972).

Appendix I. Theoretical models for industrial processes

Consider an industrial process in which a sequence of items is produced, and suppose that a quality measurement is made on the items. Let the measurements be denoted X_1, X_2, . . . , then in many cases the distribution of X will be approximately normal but with a mean $\mu(t)$ which tends to drift or change with time.

The Dodge model for an industrial process was discussed in Chapter 3, and this assumes that $\mu(t)$ is constant until changed by some assignable cause. We remarked in Chapter 3 that this model is rather unrealistic, and in many processes the $\mu(t)$ tends to wander around a central stationary value, as in Figure 3.6. A more realistic mathematical model is to put

$$Y_t = \mu(t) - \theta$$

and let Y_t be a first-order autoregressive process,

$$Y_t = \lambda Y_{t-1} + Z_t, \qquad t = 1, 2, 3, \ldots,$$

where Z_t is a random error term and λ is a constant less than unity. This process is such that the current value Y_t has two components, one arising from the current value of the process at the previous time point $(t - 1)$, and the other being a random error term. Let $E(Z_t) = 0$, $V(Z_t) = \sigma^2$, and let the Z_t's be independent.

If this model were simulated, we would see that for $\lambda < 1$, the marginal distribution Y_t tends to settle down to a stable form (normally distributed) with $E(Y_t) = 0$, $V(Y_t) = \sigma^2/(1 - \lambda^2)$. Therefore this model for $\mu(t), t = 1, 2, 3, \ldots$, leads to $\mu(t)$ having a normal distribution with $E\{\mu(t)\} = \theta$, $V\{\mu(t)\} = \sigma^2/(1 - \lambda^2)$.

The actual process for $\mu(t)$ would look something like Figure 3.6, and measurements of quality of individual items would be distributed about this mean with some other variance, say ω^2.

It is easy to check that in this model, the correlation between, say,

$\mu(t)$ and $\mu(t + n)$ is λ^n. Again this is realistic in terms of an actual process – the quality of items in a sequence tends to be correlated.

Barnard (1959) put forward a different model for an industrial process. He suggested that $\mu(t)$ be regarded as constant until it is disturbed by some event, and that the time interval between events has an exponential distribution with p.d.f. $\alpha e^{-\alpha x}$. At these events, the mean of the process is sampled from a normal distribution, with say, mean θ and variance σ^2. This model leads to a series of abrupt and irregular changes in $\mu(t)$. If desired, this kind of model could be superimposed on the autoregressive model, to simulate large abrupt changes caused by assignable causes of variation.

It is very difficult to say much in general about theoretical models for processes, as many different types of stochastic process are no doubt encountered in practice. The one point which does stand out is that as processes become more complicated, the simple Dodge model becomes more and more inadequate.

Appendix II. Statistical tables

Table 1 *Cumulative distribution function of the standard normal distribution*
(a) For x in 0·1 intervals

x	$\Phi(x)$	x	$\Phi(x)$	x	$\Phi(x)$
0·0	0·5000	1·3	0·9032	2·6	0·9953
0·1	0·5398	1·4	0·9192	2·7	0·9965
0·2	0·5793	1·5	0·9332	2·8	0·9974
0·3	0·6179	1·6	0·9452	2·9	0·9981
0·4	0·6554	1·7	0·9554	3·0	0·9987
0·5	0·6915	1·8	0·9641	3·1	0·9990
0·6	0·7257	1·9	0·9713	3·2	0·9993
0·7	0·7580	2·0	0·9772	3·3	0·9995
0·8	0·7881	2·1	0·9821	3·4	0·99966
0·9	0·8159	2·2	0·9861	3·5	0·99977
1·0	0·8413	2·3	0·9893	3·6	0·99984
1·1	0·8643	2·4	0·9918	3·7	0·99989
1·2	0·8849	2·5	0·9938	3·8	0·99993

(*b*) For x in 0·01 intervals

x	$\Phi(x)$	x	$\Phi(x)$	x	$\Phi(x)$
1·60	0·9452	1·87	0·9693	2·14	2·9838
1·61	0·9463	1·88	0·9699	2·15	0·9842
1·62	0·9474	1·89	0·9706	2·16	0·9846
1·63	0·9484	1·90	0·9713	2·17	0·9850
1·64	0·9495	1·91	0·9719	2·18	0·9854
1·65	0·9505	1·92	0·9726	2·19	0·9857
1·66	0·9515	1·93	0·9732	2·20	0·9861
1·67	0·9525	1·94	0·9738	2·21	0·9865
1·68	0·9535	1·95	0·9744	2·22	0·9868
1·69	0·9545	1·96	0·9750	2·23	0·9871
1·70	0·9554	1·97	0·9756	2·24	0·9875
1·71	0·9564	1·98	0·9761	2·25	0·9878
1·72	0·9573	1·99	0·9767	2·26	0·9881
1·73	0·9582	2·00	0·9772	2·27	0·9884
1·74	0·9591	2·01	0·9778	2·28	0·9887
1·75	0·9599	2·02	0·9783	2·29	0·9890
1·76	0·9608	2·03	0·9788	2·30	0·9893
1·77	0·9616	2·04	0·9793	2·31	0·9896
1·78	0·9625	2·05	0·9798	2·32	0·9898
1·79	0·9633	2·06	0·9803	2·33	0·9901
1·80	0·9641	2·07	0·9808	2·34	0·9904
1·81	0·9649	2·08	0·9812	2·35	0·9906
1·82	0·9656	2·09	0·9817	2·36	0·9909
1·83	0·9664	2·10	0·9821	2·37	0·9911
1·84	0·9671	2·11	0·9826	2·38	0·9913
1·85	0·9678	2·12	0·9830	2·39	0·9916
1·86	0·9686	2·13	0·9834	2·40	0·9918

Table 2 *Percentiles of the standard normal distribution*

P	x	P	x	P	x
20	0·8416	5	1·6449	2	2·0537
15	1·0364	4	1·7507	1	2·3263
10	1·2816	3	1·8808	0·5	2·5758
6	1·5548	2·5	1·9600	0·1	3·0902

Table 3 *Percentage points of the χ^2-distribution*

Degrees of freedom	*Percentage points* 1	2·5	5·0	10·0	50·0	90·0	95·0	97·5	99·0
1	$0{\cdot}0^3157$	$0{\cdot}0^3982$	$0{\cdot}0^2393$	0·0158	0·45	2·71	3·84	5·02	6·63
2	0·0201	0·0506	0·103	0·211	1·39	4·61	5·99	7·38	9·21
3	0·115	0·216	0·352	0·584	2·36	6·25	7·81	9·35	11·34
4	0·297	0·484	0·711	1·06	3·36	7·78	9·49	11·14	13·28
5	0·554	0·831	1·15	1·61	4·35	9·24	11·07	12·83	15·09
6	0·872	1·24	1·64	2·20	5·35	10·64	12·59	14·45	16·81
7	1·24	1·69	2·17	2·83	6·35	12·02	14·07	16·01	18·48
8	1·65	2·18	2·73	3·49	7·34	13·36	15·51	17·53	20·09
9	2·09	2·70	3·33	4·17	8·34	14·68	16·92	19·02	21·67
10	2·56	3·25	3·94	4·87	9·34	15·99	18·31	20·48	23·21
12	3·57	4·40	5·23	6·30	11·34	18·55	21·03	23·34	26·22
14	4·66	5·63	6·57	7·79	13·34	21·06	23·68	26·12	29·14
16	5·81	6·91	7·96	9·31	15·34	23·54	26·30	28·85	32·00
18	7·01	8·23	9·39	10·86	17·34	25·99	28·87	31·53	34·81
20	8·26	9·59	10·85	12·44	19·34	28·41	31·41	34·17	37·57
22	9·54	10·98	12·34	14·04	21·34	30·81	33·92	36·78	40·29
24	10·86	12·40	13·85	15·66	23·34	33·20	36·42	39·36	42·98
26	12·20	13·84	15·38	17·29	25·34	35·56	38·89	41·92	45·64
28	13·56	15·31	16·93	18·94	27·34	37·92	41·34	44·56	48·28
30	14·95	16·79	18·49	20·60	29·34	40·26	43·77	46·98	50·89
32	16·36	18·29	20·07	22·27	31·34	42·58	46·19	49·48	53·49
34	17·79	19·81	21·66	23·95	33·34	44·90	48·60	51·97	56·06
36	19·23	21·34	23·27	25·64	35·34	47·21	51·00	54·44	58·62
38	20·69	22·88	24·88	27·34	37·34	49·51	53·38	56·90	61·16
40	22·16	24·43	26·51	29·05	39·34	51·81	55·76	59·34	63·69
42	23·65	26·00	28·14	30·77	41·34	54·09	58·12	61·78	66·21

Table 4 *Conversion of range to standard deviation*

n	a_n	n	a_n	n	a_n	n	a_n
2	0·8862	6	0·3946	10	0·3249	14	0·2935
3	0·5908	7	0·3698	11	0·3152	15	0·2880
4	0·4857	8	0·3512	12	0·3069	16	0·2831
5	0·4299	9	0·3367	13	0·2998	17	0·2787

An estimate of σ of a normal population can be obtained by multiplying the range of a sample of size n, or the average range of a set of samples of the same size n, by a_n.

[Reproduced by permission from *Biometrika Tables for Statiticians.* Pearson and Hartley (Cambridge University Press.)]

Table 5 *Percentage points of the distribution of the relative range* (range/σ)

Sample size	0·1	1·0	2·5	5·0	90	95·0	97·5	99·0	99·9
2	0·00	0·02	0·04	0·09	2·33	2·77	3·17	3·64	4·65
3	0·06	0·19	0·30	0·43	2·90	3·31	3·68	4·12	5·06
4	0·20	0·43	0·59	0·76	3·24	3·63	3·98	4·40	5·31
5	0·37	0·66	0·85	1·03	3·48	3·86	4·20	4·60	5·48
6	0·54	0·87	1·06	1·25	3·66	4·03	4·36	4·76	5·62
7	0·69	1·05	1·25	1·44	3·81	4·17	4·49	4·88	5·73
8	0·83	1·20	1·41	1·60	3·93	4·29	4·61	4·99	5·82
9	0·96	1·34	1·55	1·74	4·04	4·39	4·70	5·08	5·90
10	1·08	1·47	1·67	1·86	4·13	4·47	4·79	5·16	5·97
11	1·20	1·58	1·78	1·97	4·21	4·55	4·86	5·23	6·04
12	1·30	1·68	1·88	2·07	4·28	4·62	4·92	5·29	6·09

Table 6 *Values of* $r(c) = \chi^2_{1-\beta}/\chi^2_\alpha$ *with d.f.* $= 2(c + 1)$

(a) $\alpha = 0{\cdot}100$	$1 - \beta$				(b) $\alpha = 0{\cdot}050$	$1 - \beta$			
c	0·900	0·950	0·975	0·990	c	0·900	0·950	0·975	0·990
0	21·85	28·43	35·01	43·71	0	44·89	58·40	71·92	89·78
1	7·31	8·92	10·48	12·48	1	10·95	13·35	15·68	18·68
2	4·83	5·71	6·56	7·63	2	6·51	7·70	8·84	10·28
3	3·83	4·44	5·02	5·76	3	4·89	5·67	6·42	7·35
4	3·29	3·76	4·21	4·77	4	4·06	4·65	5·20	5·89
5	2·94	3·34	3·70	4·16	5	3·55	4·02	4·47	5·02
6	2·70	3·04	3·35	3·74	6	3·21	3·60	3·98	4·44
7	2·53	2·82	3·10	3·44	7	2·96	3·30	3·62	4·02
8	2·39	2·66	2·90	3·20	8	2·77	3·07	3·36	3·71
9	2·28	2·52	2·75	3·02	9	2·62	2·89	3·15	3·46
10	2·19	2·42	2·62	2·87	10	2·50	2·75	2·98	3·27
11	2·12	2·33	2·51	2·74	11	2·40	2·63	2·84	3·10
12	2·06	2·25	2·42	2·64	12	2·31	2·53	2·73	2·97
13	2·00	2·18	2·35	2·55	13	2·24	2·44	2·63	2·85
14	1·95	2·12	2·28	2·47	14	2·18	2·37	2·54	2·75
15	1·91	2·07	2·22	2·40	15	2·12	2·30	2·47	2·66
16	1·87	2·03	2·17	2·34	16	2·07	2·24	2·40	2·59
17	1·84	1·99	2·12	2·29	17	2·03	2·19	2·34	2·52
18	1·81	1·95	2·08	2·24	18	1·99	2·15	2·29	2·46
19	1·78	1·92	2·04	2·19	19	1·95	2·10	2·24	2·40
20	1·76	1·89	2·01	2·15	20	1·92	2·07	2·20	2·35

(c) $\alpha = 0{\cdot}025$	$1-\beta$				(d) $\alpha = 0{\cdot}010$	$1-\beta$			
c	0·900	0·950	0·975	0·990	c	0·900	0·950	0·975	0·990
0	90·95	118·33	145·70	181·89	0	229·10	298·07	367·04	458·21
1	16·06	19·59	23·00	27·41	1	26·18	31·93	37·51	44·69
2	8·60	10·18	11·68	13·59	2	12·21	14·44	16·57	19·28
3	6·13	7·11	8·04	9·22	3	8·12	9·42	10·65	12·20
4	4·92	5·64	6·31	7·15	4	6·25	7·16	8·01	9·07
5	4·21	4·77	5·30	5·95	5	5·20	5·89	6·54	7·34
6	3·74	4·21	4·64	5·18	6	4·52	5·08	5·60	6·25
7	3·41	3·81	4·18	4·63	7	4·05	4·52	4·96	5·51
8	3·16	3·51	3·83	4·23	8	3·70	4·12	4·49	4·96
9	2·96	3·28	3·56	3·92	9	3·44	3·80	4·14	4·55
10	2·81	3·09	3·35	3·67	10	3·23	3·56	3·85	4·22
11	2·68	2·69	3·17	3·47	11	3·06	3·35	3·63	3·96
12	2·57	2·81	3·03	3·30	12	2·92	3·19	3·44	3·74
13	2·48	2·70	2·90	3·15	13	2·80	3·05	3·28	3·56
14	2·40	2·61	2·80	3·03	14	2·69	2·93	3·14	3·40
15	2·33	2·53	2·71	2·92	15	2·60	2·82	3·02	3·27
16	2·27	2·45	2·62	2·83	16	2·52	2·73	2·29	3·15
17	2·21	2·39	2·55	2·74	17	2·45	2·65	2·83	3·05
18	2·16	2·33	2·49	2·67	18	2·39	2·58	2·75	2·96
19	2·12	2·28	2·43	2·61	19	2·34	2·52	2·68	2·87
20	2·08	2·24	2·38	2·55	20	2·29	2·46	2·61	2·80

Table 7 *Factors for construction of $\bar{x}$-charts; limits at 5% and 0·2% points*

	From standard deviation		*From average range*	
Sample size	*Limit Warning*	*Action*	*Limit Warning*	*Action*
2	1·3859	2·1851	1·2282	1·9364
3	1·1316	1·7841	0·6686	1·0541
4	0·9800	1·5451	0·4760	0·7505
5	0·8765	1·3820	0·3768	0·5941
6	0·8002	1·2616	0·3157	0·4978
7	0·7408	1·1680	0·2738	0·4319
8	0·6930	1·0926	0·2434	0·3837
9	0·6533	1·0300	0·2200	0·3468
10	0·6198	0·9772	0·2014	0·3175
11	0·5910	0·9317	0·1863	0·2937
12	0·5658	0·8921	0·1736	0·2738

Table 8 *Factors for construction of $\bar{x}$-charts; limits at ± 2 and ± 3 times standard error*

	From standard deviation		*From average range*	
Sample size	*Limit Warning*	*Action*	*Limit Warning*	*Action*
2	1·4142	2·1213	1·2533	1·8799
3	1·1547	1·7321	0·6822	1·0233
4	1·0000	1·5000	0·4857	0·7286
5	0·8944	1·3416	0·3845	0·5768
6	0·8165	1·2247	0·3222	0·4833
7	0·7559	1·1339	0·2795	0·4193
8	0·7071	1·0607	0·2483	0·3725
9	0·6667	1·0000	0·2245	0·3367
10	0·6325	0·9487	0·2055	0·3082
11	0·6030	0·9045	0·1901	0·2851
12	0·5774	0·8660	0·1772	0·2658

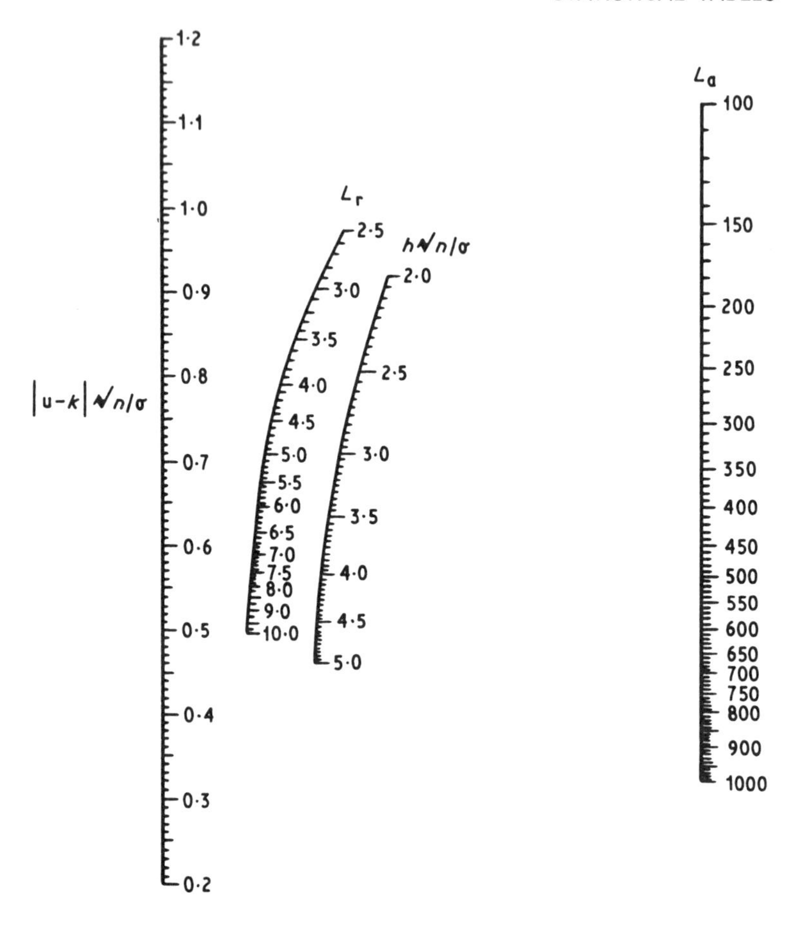

Nomogram for decision interval CUSUM schemes

L_a = ARL at AQL; L_r = ARL at RQL

[Reprinted by permission from Kemp, K. W., *Applied Statistics* (1962), **9**, 23.]

References

ANSCOMBE, F. J. (1958) 'Rectifying inspection of a continuous output', *J. Amer. Statist. Ass.*, **53,** 702–19.

ARMSTRONG, W. M. (1946) 'Foundry applications of quality control', *Industrial Quality Control,* **2** (6), 12–16.

ASF Tables (1944) *Sampling Inspection Procedures; Quality Control.* Army Services Forces, Office of the Chief of Ordnance, Washington.

BAGSHAW, M. L., and JOHNSON, N. L. (1975) 'The effect of serial correlation on the performance of CUSUM tests. II', *Technometrics,* **17,** 73–80.

BANZHAF, R. A., and BRUGER, R. M. (1970) 'MIL-STD-1235 (ORD) single and multi-level continuous sampling procedures and tables for inspection by attributes', *J. Qual. Tech.,* **2,** 41–53.

BARNARD, G. A. (1954) 'Sampling inspection and statistical decisions', *J. R. Statist. Soc., B,* **16,** 151–74.

BARNARD, G. A. (1959) 'Control charts and stochastic processes', *J. R. Statist. Soc., B,* **21,** 239–71.

BATHER, J. A. (1963) 'Control charts and the minimisation of costs', *J. R. Statist. Soc., B,* **25,** 49–80.

BEATTIE, D. W. (1962) 'A continuous acceptance sampling procedure based upon a cumulative sum chart for the number of defectives', *Applied Statistics,* **11,** 137–47.

BEATTIE, D. W. (1968) 'Patrol inspection', *Applied Statistics,* **17,** 1–16.

BEJA, A., and LADANY, S. P. (1974) 'Efficient sampling by artificial attributes', *Technometrics,* **16,** 601–11.

BISHOP, A. B. (1957) 'A model for optimum control of stochastic sampled data systems', *Operations Research,* **5,** 546–50.

BISHOP, A. B. (1960) 'Discrete random feedback models in industrial quality control', The Ohio State University Engineering Experimental Station Bull., 183v, 29n, 5th September 1960.

BISSELL, A. F. (1969) 'Cusum techniques for quality control', *Applied Statistics,* **18,** 1–30.

BOWKER, A. H. (1956) 'Continuous sampling plans', *Proc. Third Berkeley Symp. Math. Statist. & Prob.,* Univ. of California Press, Berkeley and Los Angeles. pp. 75–85.

BOX, G. E. P., and JENKINS, G. M. (1962) 'Some statistical aspects of adaptive optimisation and control', *J. R. Statist. Soc., B,* **24,** 297–343.

BOX, G. E. P., and JENKINS, G. M. (1963) 'Further contributions to adaptive quality control: Simultaneous estimation of dynamics:

Non-zero costs', Technical Report No. 19, Dept. of Statistics, Univ. of Wisconsin.

BRITISH STANDARD: 1313 (1947) *Fraction Defective Charts for Quality Control* (Dudding, B. P., and Jennett, W. J.), British Standards Institution, London.

BRITISH STANDARD: 6002 (1975) *Draft British Standard Specification Sampling Procedures and Charts for Inspection by Variables.* British Standards Institution, London.

BROOK, D. and EVANS, D. A. (1972) 'An approach to the probability distribution of cusum run length', *Biometrika,* **59,** 539–50.

CHAMPERNOWNE, D. G. (1953) 'The economics of sequential sampling procedures for defectives', *Applied Statistics,* **2,** 118–30.

CHIU, W. K. (1973) 'Comments on the economic design of x̄-charts', *J. Amer. Statist. Ass.,* **68,** 919–21.

CHIU, W. K. (1975) 'Economic design of attribute control charts', *Technometrics,* **17,** 81–7.

CHIU, W. K., and WETHERILL, G. B. (1973) 'The economic design of continuous inspection procedures: a review paper', *Inst. Stat. Rev.,* **41,** 357–73.

CHIU, W. K., and WETHERILL, G. B. (1974) 'A simplified scheme for the economic design of x̄-charts', *J. Qual. Tech.,* **6,** 63–9.

CHIU, W. K., and WETHERILL, G. B. (1975) 'Quality control practices', *Int. J. Prod. Res.,* **13,** 175–82.

COX, D. R. (1960) 'Serial sampling acceptance schemes derived from Bayes's theorem', *Technometrics,* **2,** 353–60.

DE BRUYN, C. S. VAN DOBBEN (1968) *Cumulative Sum Tests: Theory and Practice,* Griffin, London.

DEF-STAN-05/30 (1974) *Sampling Procedures and Charts for Inspection by Variables.* United Kingdom Quality Assurance Directorate, London.

DEF-131A (1966) *Sampling Procedures and Tables for Inspection by Attributes.* H.M.S.O., London.

DERMAN, C., JOHNS, M. V., JNR, and LIEBERMAN, G. J. (1959) 'Continuous sampling procedures without control', *Ann. Math. Statist.,* **30,** 1175–91.

DERMAN, C., LITTAUER, S., and SOLOMON, H. (1957) 'Tightened multilevel continuous sampling plans', *Ann. Math. Statist.,* **28,** 395–404.

DESMOND, D. J. (1954) 'Quality control on the setting of voltage regulators', *Applied Statistics,* **3,** 65–73.

DODGE, H. F. (1943) 'A sampling inspection plan for continuous production', *Ann. Math. Statist.,* **14,** 264–79.

DODGE, H. F. (1955) 'Chain sampling inspection plan', *Industrial Quality Control,* **11,** 10–13.

DODGE, H. F. (1970) 'Notes on the evolution of acceptance sampling plans, Part IV', *J. Qual Technology,* **2,** 1–8.

DODGE, H. F., and ROMIG, H. G. (1929) 'A method of sampling inspection', *Bell Syst. Tech. J.,* **8,** 613–31.

DODGE, H. F., and ROMIG, H. G. (1959) *Sampling Inspection Tables; Single and Double Sampling*, 2nd Ed. John Wiley, New York.

DODGE, H. F., and TORREY, M. N. (1951) 'Additonal continuous sampling inspection plans', *Industrial Quality Control*, **7**, 5–9.

DODGE, H. F., and TORREY, M. N. (1951) *Continuous Sampling Inspection Plans*. Monograph 1834. Bell Telephone System Technical Publications.

DUNCAN, A. J. (1956) 'The econonic design of $\bar{x}$-charts used to maintain. current control of a process', *J. Amer. Statist. Ass.*, **51**, 228–42.

DUNCAN, A. J. (1971) 'The economic design of $\bar{x}$-charts when there is a multiplicity of assignable causes', *J. Amer. Statist. Ass.*, **66**, 107–21.

DUNCAN, A. J. (1974) *Quality Control and Industrial Statistics*. Irwin, Homewood, Illinois.

EWAN, W. D. (1963) 'When and how to use Cu-Sum charts', *Technometrics*, **5**, 1–22.

EWAN, W. D., and KEMP, K. W. (1960) 'Sampling inspection of continuous processes with no autocorrelation between successive results', *Biometrika*, **47**, 363–80.

FERTIG, K. W., and MANN, N. R. (1974) 'A decision theoretic approach to defining variables sampling plans for finite lots: single sampling for exponential and Gaussian processes', *J. Amer. Statist. Ass.*, **69**, 665–71.

FORD, J. H. (1951) 'Examples of the process curve', D.I.C. Thesis, Imperial College.

FREEMAN, H. A., FRIEDMAN, H., MOSTELLER, F., and WALLIS, W. A. (1948) *Sampling Inspection*. McGraw-Hill, New York.

GIRSHICK, M. A., and RUBIN, H. (1952) 'A Bayes approach to a quality control model', *Ann. Math. Statist.*, **23**, 114–25.

GOEL, A. L., JAIN, S. C., and WU, S. M. (1968) 'An algorithm for the determination of the economic design of $\bar{x}$-charts based upon Duncan's model', *J. Amer. Statist. Ass.*, **63**, 304–20.

GOLDSMITH, P. L., and WHITFIELD, H. (1961) 'Average run lengths in cumulative chart quality control schemes', *Technometrics*, **3**, 11–20.

GRANT, E. L., and LEAVENWORTH, R. S. (1972) *Statistical Quality Control*, 4th Ed. McGraw-Hill, Tokyo.

GRIFFITHS, B., and RAO, A. G. (1964) 'An application of least cost acceptance sampling schemes', *Unternehmensforschung* (*Operations Research*), **9**, 8–17.

GUENTHER, W. C. (1971) 'On the determination of single sampling attribute plans based upon a linear cost model and a prior distribution', *Technometrics*, **13**, 483–98.

HALD, A. (1960) 'The compound hypergeometric distribution and a system of single sampling inspection plans based on prior distributions and costs', *Technometrics*, **2**, 275–340.

HALD, A. (1964) 'Single sampling inspection plans with specified acceptance probability and minimum average costs'. Institute of Mathematical Statistics, Copenhagen.

HALD, A. (1965) 'Bayesian single sampling attribute plans for discrete prior distributions', *Mat. Fys. Skr. Dan. Vid. Selsk.*, **3,** Munksgaard, Copenhagen.

HALD, A. (1967) 'The determination of single sampling attribute plans with given producer's and consumer's risk', *Technometrics,* **9,** 401–15.

HALD, A., and KOUSGAARD (1966) 'A table for solving the equation $B(c, n, p) = P$ for $c = 0(1)100$ and 15 values of P'. Institute of Mathematical Statistics, Copenhagen.

HAMAKER, H. C. (1958) 'Some basic principles of sampling inspection by attributes', *Applied Statistics,* **7,** 149–59.

HAMAKER, H. C., and VAN STRIK, R. (1955) 'The efficiency of double sampling for attributes', *J. Amer. Statist. Ass.,* **50,** 830–49.

HILL, I. D. (1956) 'Modified control limits', *Applied Statistics,* **5,** 12–19.

HILL, I. D. (1960) 'The economic incentive provided by inspection', *Applied Statistics,* **9,** 69–81.

HILL, I. D. (1962) 'Sampling inspection and defence specification DEF-131', *J. R. Statist. Soc., A,* **125,** 31–87.

HILLIER, F. S. (1961) 'New criteria for selecting continuous sampling plans', Technical Report No. 30, Applied Mathematics and Statistics Laboratories, Stanford University.

HILLIER, F. S. (1964) 'New criteria for selecting continuous sampling plans', *Technometrics,* **6,** 161–78.

HORSNELL, G. (1954) 'The determination of single sample schemes for percentage defectives', *Applied Statistics,* **3,** 150–8.

HORSNELL, G. (1957) 'Economical acceptance sampling schemes', *J. R. Statist. Soc., A,* **120,** 148–201.

HUITSON, A., and KEEN, J. (1965) *Essentials of Quality Control.* Heinemann, London.

JOHNSON, N. L. (1961) 'A simple theoretical approach to cumulative sum charts', *J. Amer. Stat. Assoc.,* **56,** 835–40.

JOHNSON, N. L. (1966) 'Cumulative sum charts and the Weibull distribution', *Technometrics,* **8,** 481–91.

JOHNSON, N. L., and BAGSHAW, M. L. (1974) 'The effect of serial correlation on the performance of CUSUM tests', *Technometrics,* **16,** 103–12.

KAO, E. P. C. (1972) 'Economic screening of a continuously manufactured product', *Technometrics,* **14,** 653–61.

KAO, J. H. K. (1971) 'MIL-STD-414 sampling procedures and tables for inspection by variables for percent defective', *J. Qual. Tech.,* **3,** 28–37.

KEMP, K. W. (1958) 'Formulae for calculating the operating characteristic and the average sample number of some sequential tests', *J. R. Statist. Soc., B,* **20,** 379–86.

KEMP, K. W. (1961) 'The average run length of the cumulative sum chart when a V-mask is used', *J. R. Statist. Soc., B,* **23,** 149–53.

KEMP, K. W. (1962) 'The use of cumulative sums for sampling inspection schemes', *Applied Statistics,* **11,** 16–31.

KEMP, K. W. (1971) 'Formal expressions which can be applied to CUSUM charts', *J. R. Statist. Soc., B*, **33,** 331–60.

KING, E. P. (1952) 'The operating characteristic of the control chart for sample means', *Ann. Math. Statist.*, **23,** 384–95.

LAI, T. L. (1974) 'Control charts based on weighted sums', *Ann. Statist.*, **2,** 134–47.

LIEBERMAN, G. J. (1953) 'A note on Dodge's continuous inspection plan', *Ann. Math. Statist.*, **24,** 480–4.

LIEBERMAN, G. J. (1965) 'Statistical process control and the impact of automatic process control', *Technometrics*, **7,** 283–92.

LIEBERMAN, G. J. and RESNIKOFF, G. J. (1955) 'Sampling plans for inspection by variables', *J. Amer. Statist. Ass.*, **50,** 457–516 and 1333.

LIEBERMAN, G. J. and SOLOMON, H. (1954) 'Multi-level continuous sampling plans', Technical Report No. 17, Applied Mathematics and Statistics Laboratory, Stanford University.

LIEBERMAN, G. J. and SOLOMON, H. (1955) 'Multi-level continuous sampling plans', *Ann. Math. Statist.*, **26,** 686–704.

LUCAS, J. M. (1973) 'A modified V-mask control scheme', *Technometrics*, **15,** 833–7.

MIL-STD105D (1963) Department of Defence, U.S. Government Printing Office, Washington, D.C.

MIL-STD-414, (1957) *Sampling Procedures and Tables for Inspection by Variables for Percent Defective.* U.S. Government Printing Office, Washington, D.C.

MIL-STD-1235 (ORD) (1962) *Single and Multi-level Continuous Sampling Procedures and Tables for Inspection by Attributes.* Dept. of the Army, Washington, D.C.

MOOD, A. M. (1943) 'On the dependence of sampling inspection plans upon population distributions', *Ann. Math. Statist.*, **14,** 415–25.

MORGAN, M. E., MACLEOD, P., ANDERSON, E. O., and BLISS, C. I. (1951) 'A sequential procedure for grading milk by microscopic counts', *Conn. (Storrs) Agric. Exp. Sta., Bull.*, 276, 35 pp.

OWEN, D. B. (1964) 'Control of percentages in both tails of the normal distribution', *Technometrics*, **6,** 377–88.

OWEN, D. B. (1967) 'Variables sampling plans based on the normal distribution', *Technometrics*, **9,** 417–23.

OWEN, D. B. (1968) 'A survey of properties and applications of the non-central t-distribution', *Technometrics*, **10,** 455–78.

OWEN, D. B. (1969) 'Summary of recent work on variables acceptance sampling with emphasis on non-normality', *Technometrics*, **11,** 631–7.

PAGE, E. S. (1954) 'Continuous inspection schemes', *Biometrika*, **41,** 100–15.

PAGE, E. S. (1955) 'Control charts with warning lines', *Biometrika*, **42,** 243–57.

PAGE, E. S. (1961) 'Cumulative sum charts', *Technometrics*, **3,** 1–9.

PAGE, E. S. (1962) 'Cumulative sum charts using gauging', *Technometrics*, **4,** 97–109.

PEACH, P. (1947) *Industrial Statistics and Quality Control.* 2nd ed., Raleigh, N. C. Edwards & Broughton.

PFANZAGL, J. (1963) 'Sampling procedures based on prior distributions and costs', *Technometrics,* **5,** 47–61.

PHILLIPS, M. J. (1969) 'A survey of sampling procedures for continuous production', *J. R. Statist. Soc., A,* **132,** 100.

PRAIRIE, R. R., and ZIMMER, W. J. (1970) 'Continuous sampling plans based on cumulative sums', *Applied Statistics,* **19,** 222–30.

RAI, GULAB (1971) 'Continuous acceptance sampling procedure based on cumulative sum chart for mean', *Trabajos de Estadistica, Investigacion Operativa,* **22,** 213–19.

READ, D. R., and BEATTIE, D. W. (1961) 'The variable lot-size acceptance sampling plan for continuous production', *Applied Statistics,* **10,** 147–56.

RESNIKOFF, G. J., and LIEBERMAN, G. J. (1957) *Tables of the Non-Central t-distribution.* Stanford Univ. Press, Stanford, Cal.

ROBERTS, S. W. (1958) 'Properties of central chart zone tests', *Bell System Technical Journal,* **37,** 83–114.

ROBERTS, S. W. (1959) 'Control chart tests based on geometric moving averages', *Technometrics,* **1,** 239–50.

ROBERTS, S. W. (1966) 'A comparison of some control chart procedures', *Technometrics,* **8,** 411–30.

SACKROWITZ, H. (1972) 'Alternative multi-level continuous sampling plans', *Technometrics,* **14,** 645–52.

SAVAGE, I. R. (1959) 'A production model and continuous sampling plan', *J. Amer. Statist. Ass.,* **54,** 231–47.

SCHMIDT, J. W., and TAYLOR, R. E. (1973) 'A dual purpose cost based quality control system', *Technometrics,* **15,** 151–66.

SITTIG, J. (1951) 'The economic choice of sampling system in acceptance sampling', *Bull. Int. Statist. Inst.,* **33,** V, 51–84.

TAYLOR, H. M. (1968) 'The economic design of cumulative sum control charts', *Technometrics,* **10,** 479–88.

TRAUX, H. M. (1961) 'Cumulative sum charts and their application to the Chemical Industry', *Industrial Quality Control,* no. **18,** 6 (Milwaukee).

WALD, A. (1947) *Sequential Analysis.* John Wiley, New York.

WETHERILL, G. B. (1959) 'The most economical sequential sampling scheme for inspection by variables', *J. R. Statist. Soc.,* B, **21,** 400–8.

WETHERILL, G. B. (1960) 'Some remarks on the Bayesian solution of the single sample inspection scheme', *J. R. Statist. Soc., B,* **25,** 1–48.

WETHERILL, G. B. (1975) *Sequential Methods in Statistics.* Chapman and Hall, London.

WETHERILL, G. B., and CAMPLING, G. E. G. (1966) 'The decision theory approach to sampling inspection', *J. R. Statist. Soc., B,* **28,** 381–416.

WETHERILL, G. B., and CHIU, W. K. (1974) 'A simplified attribute sampling scheme', *Applied Statistics,* **23,** 143–8.

WETHERILL, G. B., and CHIU, W. K. (1975) 'A review of acceptance sampling schemes with emphasis on the economic aspect', *Int. Stat. Rev.*, **43,** 191–209.

WHITE, L. S. (1965) 'Markovian decision models for the evaluation of a large class of continuous sampling inspection plans'. *Ann. Math. Statist.*, **36,** 1408–20.

WILSON, E. B., and BURGESS, A. R. (1971) 'Multiple sampling plans viewed as finite Markov chains', *Technometrics*, **13,** 371–82.

WOODWARD, R. H., and GOLDSMITH, P. L. (1964) *Cumulative Sum Techniques*. I.C.I. Monograph No. 3. Oliver & Boyd, Edinburgh.

Index